Head-Related Transfer Function and Acoustic Virtual Reality

Kazuhiro Iida

Head-Related Transfer Function and Acoustic Virtual Reality

Springer

Kazuhiro Iida
Spatial Hearing Laboratory, Faculty of Advanced Engineering
Chiba Institute of Technology
Narashino, Chiba, Japan

Original Japanese edition published by Corona-Sha, Tokyo, 2017.

ISBN 978-981-13-9744-8 ISBN 978-981-13-9745-5 (eBook)
https://doi.org/10.1007/978-981-13-9745-5

This Springer imprint is published by the registered company Springer Nature Singapore Pte Ltd.
The registered company address is: 152 Beach Road, #21-01/04 Gateway East, Singapore 189721, Singapore

Preface

This book describes the latest findings from the fundamentals to applications of the head-related transfer function (HRTF). The HRTF plays a primary role in human perception of the direction of sound. By applying HRTFs, the three-dimensional acoustical sensation of an existing sound field (at concert halls, stadiums, or disaster sites) can be reproduced, and/or that of an imaginary sound field can be generated for video games and cinemas, independent of time and space. Such systems are currently being developed.

Starting from the 1960s, when studies on HRTFs became active, a number of researchers around the world have obtained various findings. Considerable fundamental knowledge on HRTFs has been accumulated. However, no major breakthrough has been found in the application of HRTFs. At present, a truly three-dimensional acoustic VR system for practical consumer use does not exist anywhere in the world. One of the principal reasons for this is that the problem on individual differences in HRTFs has not been solved.

Blauert's scenario was introduced in March 1999: "*A person who enters a multimedia shop is scanned by a camera and some instants later his/her individual HRTF set is ready to be sold for the use in advanced 3D applications*".

Although 20 years have passed since then, this scenario has not yet been realized. The current acoustic VR systems provide a three-dimensional acoustical sensation only for a specific listener. Studies on individualization methods for HRTFs have been actively performed. (There are many challenges throughout the world besides the research described in this book). In the future, an HRTF suitable for each listener will be generated from images of the head and pinna of the listener captured by the camera of a smart phone and will be delivered to the listener via the Internet. This process would be even simpler than buying glasses.

Three-dimensional acoustic VR has the potential to improve society and life in a wide range of fields, not only in entertainment, but also in highly specialized education and training, research on human perception and recognition, high-precision control of robots and devices, architecture and urban design, immersive communication, and new artistic expressions. Based on such social demands, I expect that an acoustic VR system, which provide a three-dimensional acoustical sensation for anyone can be realized within 10 years.

This book is a revised version of the "Fundamentals of head-related transfer functions and its application to the 3-D sound system" written in Japanese by me as part of the acoustic technology series of the Acoustical Society of Japan. I would like to thank the Japan Acoustical Society and the Corona Publishing Co. Ltd. for their permission to revise and republish the original book.

Mr. Yasutaka Okazaki, Springer Japan, put forward the publication plan for this book. Ms. Emmy Lee and Ms. Kripa Guruprasad, Springer Nature, patiently awaited completion of the manuscript and published this book. My laboratory secretary Ms. Naoko Tanaka prepared the reference list and helped with the proofreading. In addition to the university laboratory and my home study, pelican coffee in Den-en-chofu, Tokyo provided a third writing place. My family and my dog Pluto gave me the enthusiasm to go on.

While I appreciate their cooperation, if there is an error in the text of this book, it is attributable solely to myself. If you have notes, I would appreciate your comments (e-mail: kiida.hrtf@gmail.com).

Narashino, Japan Kazuhiro Iida
May 2019

Contents

Chapter 1
Introduction

Abstract This introductory chapter describes the motivation for writing this book. Then, head-related transfer functions (HRTFs) are defined, and the coordinate system used in this book is described. Finally, recent advancements in HRTF research are briefly summarized.

1.1 Why Research HRTFs

Humans perceive the direction and distance of a sound source in three-dimensional space with only two ears. Geometrically, n + 1 receiving points are required to identify the position of an object in n-dimensional space. Limiting the discussion to the direction of sound, it is thus impossible to identify the direction geodetically with two receiving points (two ears). So, how do we perceive the direction of sound? This is the starting point of HRTF research.

An easy way to identify the direction of sound is to rotate your head. If you rotate your head so that you perceive a sound as being in front of you, you are now looking in the direction of the sound source. However, humans can perceive the direction of sound without rotating their heads, and in fact rarely rotate their heads spontaneously to identify the direction of sound (Nojima et al. 2013). Even barn owls, which catch prey by identifying the direction of the sound, do not use head movement in performing this task (Knudsen and Konishi 1979).

HRTFs play a key role in the perception of sound direction. With few exceptions (e.g., listening to music through headphones and telephone conversations), most of the sounds we hear in everyday life are affected by the HRTF. Therefore, research on HRTFs is absolutely essential to understanding the mechanism behind the perception of sound direction.

On the other hand, the reproduction of an existing sound field and/or the creation of a virtual sound field beyond time and space, i.e., acoustic virtual reality (VR) and augmented reality (AR), can be achieved through the use of HRTFs.

The original version of this chapter was revised. The correction to this chapter is available at https://doi.org/10.1007/978-981-13-9745-5_14

K. Iida, *Head-Related Transfer Function and Acoustic Virtual Reality*,
https://doi.org/10.1007/978-981-13-9745-5_1

As such, HRTFs play an important role in both research on spatial hearing and development of acoustical VR systems. So, what exactly is an HRTF?

1.2 What Is an HRTF?

Sound waves are affected by the head, pinnae, and torso of the listener before reaching the listener's ear drums. The HRTF is the physical change in the sound wave expressed in the frequency domain due to these effects. Figure 1.1 shows an amplitude spectrum of an HRTF for a sound source located in front of a subject. Here, 0 dB indicates the amplitude of the sound pressure measured without the head, pinnae, and torso. There exist several distinguishing peaks and notches that exceed ± 10 dB. A positive value of the amplitude of the HRTF indicates that the sound pressure increases as a result of the head, pinnae, and torso of the subject, whereas a negative value indicates that the sound pressure decreases as a result of these components. Humans hear such a sound with spectral peaks and notches in everyday life.

The HRTF is defined as follows:

$$H_{1,\mathrm{r}}(s, \alpha, \beta, r, \omega) = \frac{G_{1,\mathrm{r}}(s, \alpha, \beta, r, \omega)}{F(\alpha, \beta, r, \omega)} \tag{1.1}$$

where $G_{l,r}$ is the transfer function between a sound source and the entrance of the ear canal[1] of the subject in the free field, F is the transfer function between a sound

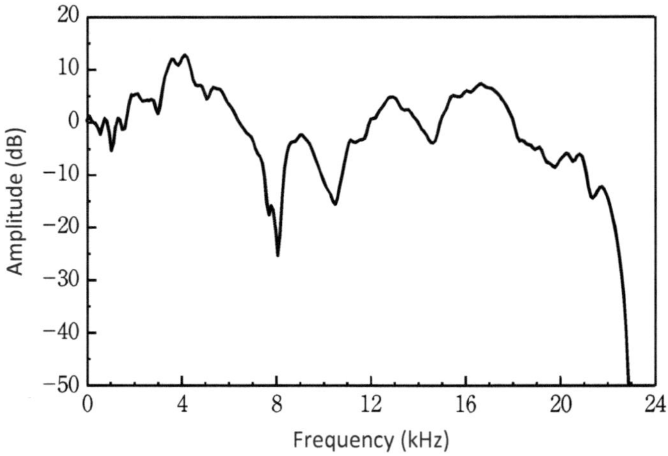

Fig. 1.1 Example of the amplitude of an HRTF for the forward direction

[1]There is another way to define the HRTF, in which the receiving point is the ear drum rather than the entrance of the ear canal (Wightman and Kistler 1989a, b). However, it is not easy to place a miniature microphone just before the ear drum to measure the HRTF. Moreover, the transfer function of the ear canal does not depend on the direction of the sound source (Møller 1992). As such, most of studies measure HRTFs using a miniature microphone located at the entrance of the ear canal.

source and the point corresponding to the center of the subject's head in the free field without the subject. The subscripts l and r denote the left and right ears, and *s* denotes the subject. Moreover, α, β, and γ denote the lateral angle, the vertical (rising) angle, and the distance, respectively. However, the distance of a sound source does not affect the HRTF when the distance exceeds 1 m (Morimoto et al. 1976).

Note that the phase spectrum of an HRTF is not described in this book because the absolute phase of the HRTF does not play an important role in the perception of sound direction (Kulkarni et al. 1999) if only the relative phase difference between the left and right ears is preserved.

The HRTF varies with the direction of a sound source. This is due to the asymmetrical shape of the head and pinnae. Humans perceive the direction of sound using the directional dependence of HRTF. The directional dependence of HRTF is described in detail in Chaps. 2 and 3.

The HRTF also varies with the listener, even if the direction of a sound source is constant, due to individual differences in shape of the head and pinna. The individualized nature of the HRTF is a serious problem in the design of practical acoustic VR systems. The details of the problem and some approaches to solving this problem are presented in Chap. 4.

1.3 HRTF and HRIR

As described above, the HRTF expresses the change in sound pressure due to the head, pinna, and torso in the frequency domain. However, the expression in the time domain may sometime be better for clarifying the characteristics of this change. The expression in the time domain is called the head-related impulse response (HRIR).

In most situations, the HRTF is obtained by Fourier transform of the measured HRIR, as follows:

$$H_{\mathrm{l,r}}(s,\alpha,\beta,r,\omega) = \frac{\mathcal{F}\left[g_{\mathrm{l,r}}(s,\alpha,\beta,r,t)\right]}{\mathcal{F}[f(\alpha,\beta,r,t)]} \tag{1.2}$$

where $g_{l,r}$ is the impulse response between a sound source and the entrance of the ear canal of the subject in the free field, *f* is the impulse response between a sound source and the point corresponding to the center of the subject's head in the free field without a subject. Moreover, $\mathcal{F}$ denotes the Fourier transform.

Figure 1.2 shows an HRIR for a sound source located in front of a subject. The response converges within approximately 2–3 ms.

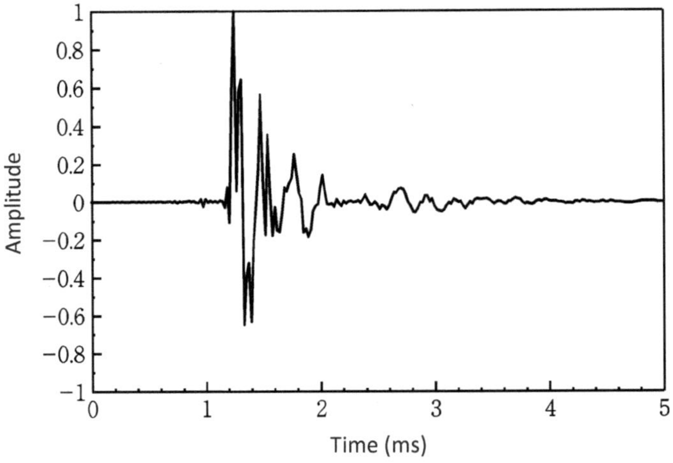

Fig. 1.2 Example of an HRIR for the forward direction

1.4 Sound Source and Sound Image

A sound wave causes many kinds of sensations when it arrives at the ear drum of a listener. What is perceived auditorily is called a sound image, or an auditory event. Whereas a sound source is a physical presence, a sound image is a psychological presence (perceptual phenomenon). A sound image includes temporal characteristics (sense of reverberation, rhythm, duration, and so on), spatial characteristics (sense of direction, distance, broadening, and so on), and qualitative characteristics (loudness, pitch, timbre, and so on). Perception of the direction and distance of a sound image is called sound image localization.

In most situations in our everyday life, a sound image is localized at the position of a sound source. However, this is not always the case. For a narrow-band signal, a sound image is sometimes localized to the front for a sound source that is actually located to the rear, and vice versa. This phenomenon is referred to as front-back confusion. Moreover, for certain narrow-band frequency ranges, a sound image is often localized in a particular direction, regardless of the direction of the actual sound source (Blauert 1969/70). These phenomena are discussed in detail in Chaps. 3, 6, and 7.

On the other hand, acoustic VR and AR systems intentionally make the listener perceive a sound image at an arbitrary position in three-dimensional space, while the sound sources (headphones) actually exist near the listener's ears.

1.5 Coordinate System

In this section, I explain the coordinate system used in this book. Figure 1.3 shows an illustration of a spherical coordinate system, which uses azimuth angle ϕ and elevation angle θ. This coordinate system is likely familiar to most readers. The azimuth angle and the elevation angle, however, do not correspond to the human mechanisms of perception of sound direction.

Thus, the interaural-polar-axis coordinate system is used in this book (Fig. 1.4). In this coordinate system, the direction of a sound source is expressed in terms of lateral angle α and vertical (rising) angle β. The lateral angle α corresponds to the mechanism for left-right perception (lateral direction) of a sound image, and the vertical angle β corresponds to that for front-back and up-down perception (vertical direction) of a sound image.

The lateral angle α is the complement of the angle between the aural axis and the line segment connecting the sound source and the origin. The origin is the center of the subject's head, and the aural axis is the straight line connecting the entrances of the left and right ear canals and the origin. The vertical angle β is the elevation within the sagittal plane, which passes through a sound source.

The horizontal plane is the plane which passes the right eye orbit and the left and right tragi (Frankfurt horizontal plane). The transverse plane is the plane perpendicular to the horizontal plane that passes through the entrances of the left and right ear canals. The median plane is the plane perpendicular to both the horizontal plane and

Fig. 1.3 Spherical coordinate system

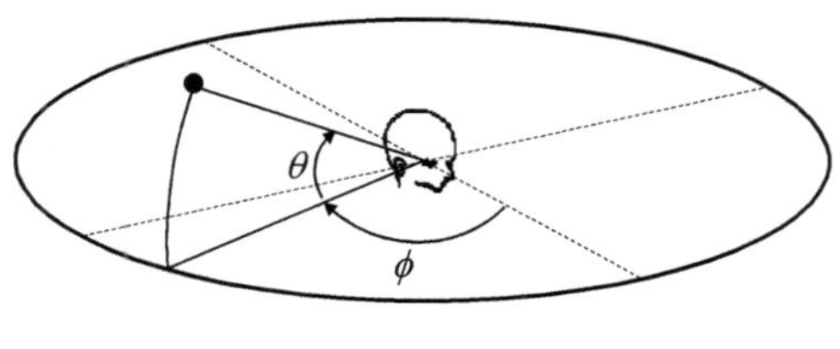

Fig. 1.4 Interaural-polar-axis coordinate system

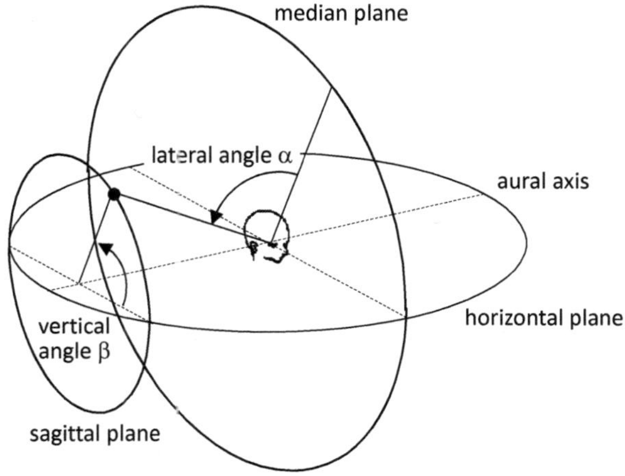

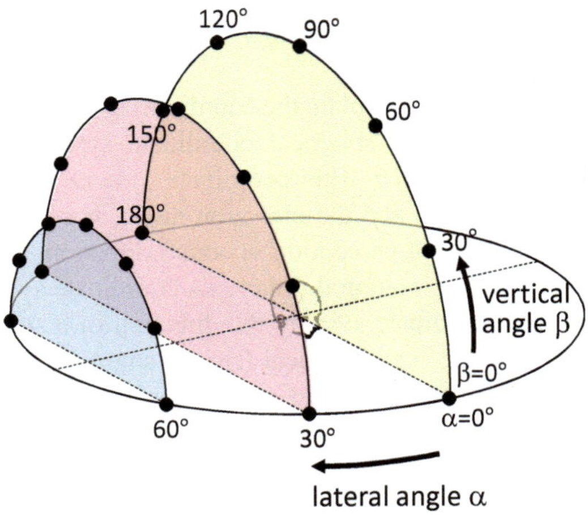

Fig. 1.5 Examples of points in sagittal planes

the transverse plane that divides the body exactly into left and right sides. The sagittal planes are the planes parallel to the median plane.

Figure 1.5 shows examples of the points, the lateral angles of which are 0°, 30°, and 60° and the vertical angles of which are 0° to 180°, in 30° steps. The interaural-polar-axis coordinate system defines the direction of a sound source using two pieces of information: the lateral sagittal plane on which the sound source is located and the elevation angle in this sagittal plane. The reader can easily understand this by, for example, imaging slices of a lemon. The lateral angle indicates the slice in which the sound source is located, and the vertical angle indicates the angle within the slice at which the sound source is located.

In the horizontal plane, however, it is easier to express the direction using the azimuth angle (0°–360°) than to use the combination of the lateral and vertical angles. For example, the expression using the combination of the lateral and vertical angles for the seven directions in the right half of the horizontal plane shown in Fig. 1.5 are (0°, 0°), (30°, 0°), (60°, 0°), (90°, 0°), (60°, 180°), (30°, 180°), and (0°, 180°). These seven directions can be expressed simply using the azimuth angle as (0°, 0°), (30°, 0°), (60°, 0°), (90°, 0°), (120°, 0°), (150°, 0°), and (180°, 0°), respectively. Therefore, in this book, the direction in the horizontal plane is expressed using the azimuth angle.

The azimuth angle ϕ and the elevation angle θ can be transformed into the lateral angle α and the vertical angle β, as follows:

$$\alpha = 90 - \cos^{-1}(\sin\phi\cos\theta)\ (^\circ) \tag{1.3}$$

$$\beta = \sin^{-1}\frac{\sin\theta}{\sqrt{\sin^2\theta + \cos^2\phi\cos^2\theta}}\ (^\circ) \tag{1.4}$$

For instance, the lateral and vertical angles for an azimuth angle of 30° and an elevation angle 45° are 20.7° and 49.1°, respectively.

1.6 Brief History of HRTF Research – Current Achievements and Research Questions to Be Settled

Although the early studies on the HRTF date back to the 1940s, research on HRTFs became active in the 1960s and has progressed by leaps and bounds in the last 60 years. Toward the end of this chapter, I summarize current achievements in research into HRTFs and open questions regarding HRTFs.

1.6.1 Origin of the HRTF

When was the concept of the HRTF proposed? As far as I know, the earliest paper on the HRTF appears to have been written by Wiener and Ross (1946). They measured HRTFs in the horizontal plane in steps of 45° for the frequency range of 200–6000 Hz. However, they did not use the term "HRTF", but rather "diffraction around the human head". The first paper to mention the term "HRTF" appears to have been written by Morimoto and Ando (1980).

1.6.2 Physical Characteristics of the HRTF

As mentioned in Sect. 1.1, there exist several distinguishing spectral peaks and notches, which exceed ± 10 dB, in the HRTF for the forward direction. For a sound source located in the lateral direction, the spectrum of the HRTF of the ears contralateral to the sound source is flattened, and the level difference between the left and right ears increases. The level difference exceeds 30 dB at particular frequency for a sound source located at an azimuth angle of 90°. The frequencies of spectral notches become higher as a sound source moves from the front to above. However, the frequencies of spectral peaks are approximately constant, regardless of the direction of the sound source. Moreover, there exist significant individual differences in the notch frequency due to individual differences in the shape and size of the pinna (Raykar et al. 2005; Iida et al. 2014).

The notches and peaks are generated in the pinna. At the notch frequency, the anti-nodes are generated at the cymba of concha and the triangular fossa, and a node is generated at the cavity of concha (Takemoto et al. 2012). Peaks are considered to be generated by the resonances of the pinna (Shaw 1997). When the cavities of the pinna are occluded by clay, the notches and peaks vanish (Iida et al. 1998), and the front-back confusion of a sound image increases (Gardner and Gardner 1973; Musicant and Butler 1984; Iida et al. 1998).

1.6.3 *Reproduction of the Direction of a Sound Image by Reproduction of the HRTF*

Morimoto and Ando (1980) demonstrated that the direction of a sound image can be reproduced by reproducing the listener's own HRTF, using a transaural system (see Sect. 12.2 for details). They also demonstrated that the HRTFs of other listeners often cause front-back confusion of a sound image and inside-of-head localization. Wightman and Kistler (1989a, b) reported that they obtained similar results using headphones with a compensation filter for the transfer function between the headphones and the ear drum.

1.6.4 *Cues for the Perception of Lateral Direction*

Around 1900, it was known that the interaural time difference and the interaural level difference are cues for the perception of lateral direction (Lord Rayleigh 1877, 1907). Afterwards, around 1960, the quantitative relationships between these interaural differences and the left-right direction of a sound image were reported. Specifically, a sound image is localized at an azimuth angle of 90° for an interaural time difference of 1 ms or an interaural level difference of 10 dB for broad-band noise (Sayers 1964; Toole and Sayers 1965).

1.6.5 *Cues for the Perception of Vertical Direction*

After the 1970s, a number of studies examined cues for the perception of vertical direction. Consequently, it was found that cues exist in the amplitude spectrum of the HRTF (referred to as spectral cue). More specifically, spectral notches and peaks above 5 kHz were found to play an important role in the perception of vertical direction (Butler and Belendiuk 1977; Musicant and Butler 1984; Hebrank and Wright 1974; Morimoto and Saito 1977; Mehrgardt and Mellert 1977), and the outline of the amplitude spectrum was found to be more important than its fine structure (Asano et al. 1990; Middlebrooks 1992, 1999; Kulkarni and Colburn 1998; Langendijk and Bronkhorst 2002; Macpherson and Sabin 2013). Hebrank and Wright (1974) first mentioned that some particular notches and peaks are more important than others in median plane localization. However, the spectral cues they considered were derived from localization tests using a narrow-band signal, and their hypothesis was found to be incorrect for a broad-band signal (Iida and Ishii 2018). On the other hand, based on sound localization tests using the parametric HRTFs, another hypothesis was proposed, in which the two lowest frequency notches (N1 and N2) above 4 kHz are the important cues (Iida et al. 2007; Iida and Ishii 2018).

Some researchers have reported that the rotation of the head is a cue for front-back discrimination of a sound source (Thurlow and Runge 1967; Perrett and Noble 1997; Kato et al. 2003). For instance, if a sound image moves to the left when you rotate the head to the right, the sound source should be in front of you. However, humans rarely rotate the head spontaneously to identify the direction of a sound source (Nojima et al. 2013). Thus, it is unlikely that humans use head rotation as a cue for front-back discrimination.

1.6.6 Physiological Mechanism for the Perception of the Direction of a Sound Image

It has been reported that there exist functions for calculating the time and level differences between the input signals to the left and right ears, in the medial and lateral nuclei of the superior olive, respectively. The superior olive is located on the pathway from the inner ear to the primary auditory cortex. Furthermore, the dorsal cochlea nucleus (DCN) extracts the spectral notch, and the DCN type IV neurons encode rising spectral edges of the notch (Reiss and Young 2005).

1.6.7 HRTF Models

Research on mathematical and physical models of the HRTF is in progress. For example, the principal component analysis (PCA) model (Kistler and Wightman 1992; Middlebrooks and Green 1992) expresses the amplitude spectrum of an HRTF by superposition of principal components (functions of the frequency domain). Although the PCA model is a beautiful mathematical model, a number of unsolved issues hinder its practical application.

A parametric HRTF model, which focuses on spectral notches and peaks, has also been proposed (Iida et al. 2007). The parametric HRTF is recomposed of some of or all of the spectral notches and peaks extracted from a listener's measured HRTF, regarding the peak around 4 kHz, which is independent of the vertical angle of the sound source (Shaw and Teranishi 1968), as the lower-frequency limit. The notches and peaks are labeled in order of frequency (e.g., P1, N1, P2, N2). The notches and peaks are expressed parametrically in terms of center frequency, level, and sharpness. They carried out sound localization tests in the upper median plane and demonstrated that the parametric HRTF recomposed of only N1, N2, P1, and P2 provided approximately the same localization performance as the measured HRTFs (Iida et al. 2007; Iida and Ishii 2018).

1.6.8 Standardization of the HRTF

There exist remarkable individual differences in HRTFs. This is a serious problem, which prevents acoustic VR from coming into widespread practical use. Current acoustic VR, which can present three-dimensional acoustical sensation to only a specific listener, must be evolved into a universal system that can present three-dimensional acoustical sensation to everyone.

In order to address the individual differences in the HRTFs of different listeners, the standardization of the HRTF has been investigated. A number of dummy-head systems have been developed as hardware solutions. For example, the Knowles Electric Manikin for Acoustic Research (KEMAR) dummy head was developed based on median values measured for dimensions of the head, pinna, and torso of many adults (Burkhard and Sachs 1975). Many of the listeners, however, confuse the front-back direction and/or perceive a sound image inside of the head when they hear sounds using the HRTFs of the KEMAR. This indicates that, at present, the standardization of the HRTF does not solve the problem of individual differences in HRTFs.

1.6.9 Individualization of the HRTF

Another solution to the problem of individual differences in HRTFs is the individualization of the HRTF, which provides personalized HRTFs suitable for each listener. The following approaches have been proposed as individualization methods of the amplitude spectrum of the HRTF:

1. Select the HRTF of the pinna that most closely resembles the pinna shape of the listener from an HRTF database (Zotkin et al. 2003).
2. Expand or compress the amplitude spectrum of the reference HRTF in the frequency domain to minimize the spectral difference between the listener's HRTF and the reference HRTF (Middlebrooks 1999).
3. Synthesize the HRTF using principal components based on the listener's pinna shape (Kistler and Wightman 1992; Middlebrooks and Green 1992).
4. Estimate the notch frequencies of the listener's HRTF from the anthropometry of the listener's pinna, and select the best matching HRTF, the notch frequencies of which are the most similar to the estimated notch frequencies, from an HRTF database (Iida et al. 2014).
5. Select a suitable HRTF from an HRTF database based on a listening test (Seeber and Fastl 2003; Iwaya 2006).

However, problems associated with practical use remain for each method. The hope is that these problems will be resolved in future studies.

1.6.10 Measurement of the HRTF

At this moment, the most reliable method by which to obtain an accurate HRTF is measurement in an anechoic chamber using Eqs. (1.1) or (1.2). However, there are two serious problems in HRTF measurement.

The first problem is related to the need to measure HRTFs with a high SN ratio. In order to accomplish this task, the types of signals suitable for measurement have been investigated. As a result, the maximum-length sequence (MLS) signal and the swept-sine signal were developed. At present, the swept-sine signal has gained mainstream acceptance.

The other problem is related to the need to measure HRTFs for various directions in a short time. A method using reciprocity, which switches the point of the sound source and the point of the receiver, has been proposed (Zotkin et al. 2006). In principle, the HRTFs of many directions can be measured simultaneously by placing a small actuator at the entrance of the ear canal and placing a number of microphones at various directions. However, improvement of the SN ratios of the measured HRTF is still necessary.

1.6.11 Numerical Calculation of the HRTF

In the twenty-first century, numerical calculation of HRTFs by the boundary element method (BEM) and the finite-difference time-domain (FDTD) method became possible using a three-dimensional wireframe model of the head and pinna (Katz 2001; Otani and Ise 2003; Kahana and Nelson 2006; Xiao and Liua 2003; Mokhtari et al. 2007). Using such numerical calculation methods, the understanding of the generation mechanism of the HRTF in the pinna has progressed. However, special equipment, e.g., magnetic resonance imaging (MRI) equipment, is required in order to create a three-dimensional wireframe model of the head and pinnae. Namely, at the moment, calculation of the HRTF is available only to a limited number of listeners. The development of an easy method by which to model the head and pinnae is necessary for the calculation of HRTFs for general listeners.

1.6.12 Directional Band

Blauert (1969/70) reported that there exist particular narrow bands (1/3 octave bands), for which a listener localizes a sound image to a specific vertical angle, regardless of the actual vertical angle of a sound source. He called these bands the directional bands. The center frequencies of the directional bands for the forward and upward directions are 4 kHz and 8 kHz, respectively. The center frequencies for the rearward direction are 1.25 kHz, 10 kHz, and 12.5 kHz. Furthermore, he reported that the band levels of the HRTF of the direction at which the directional bands occurred were higher than the band levels of the HRTFs of other directions. Blauert called these bands the boosted bands. This infers that the spectral peak dominates the perception of vertical direction for a narrow-band signal.

On the other hand, for a wide-band signal, the listener does not localize a sound image to the direction of the directional band, even if the energy of the band corresponding to the directional band is enhanced. For example, when the energy of the band, corresponding to the upward directional band (1/3 octave band of 8 kHz), of a wide-band signal was enhanced and emitted from the forward direction, the listener perceives a single sound image to the front for an enhancement of less than approximately 18 dB. For an enhancement of more than 18 dB, the listener perceives two split sound images. A sound image for the enhanced band is localized above the listener, and another broad-band sound image is localized to the front of the listener.

References

Asano F, Suzuki Y, Sone T (1990) Role of spectral cues in median plane localization. J Acoust Soc Am 88:159–168

Blauert J (1969/70) Sound localization in the median plane. Acust 22: 205–213

Burkhard MD, Sachs RM (1975) Anthropometric manikin for acoustic research. J Acoust Soc Am 58:214–222

Butler A, Belendiuk K (1977) Spectral cues utilized in the localization of sound in the median sagittal plane. J Acoust Soc Am 61:1264–1269

Gardner MB, Gardner RS (1973) Problem of localization in the median plane: effect of pinnae cavity occlusion. J Acoust Soc Am 53:400–408

Hebrank J, Wright D (1974) Spectral cues used in the localization of sound sources on the median plane. J Acoust Soc Am 56:1829–1834

Iida K, Ishii Y (2018) Effects of adding a spectral peak generated by the second pinna resonance to a parametric model of head-related transfer functions on upper median plane sound localization. Appl Acoust 129:239–247

Iida K, Yairi M, Morimoto M (1998) Role of pinna cavities in median plane localization. In: Proceedings of 16th international congress on acoustics. Acoustical Society of America, Seattle, pp 845–846

Iida K, Itoh M, Itagaki A, Morimoto M (2007) Median plane localization using parametric model of the head-related transfer function based on spectral cues. Appl Acoust 68:835–850

Iida K, Ishii Y, Nishioka S (2014) Personalization of head-related transfer functions in the median plane based on the anthropometry of the listener's pinnae. J Acoust Soc Am 136:317–333

Iwaya Y (2006) Individualization of head–related transfer functions with tournament–style listening test: listening with other's ears. Acoust Sci Tech 27:340–343

Kahana Y, Nelson PA (2006) Numerical modelling of the spatial acoustic response of the human pinna. J Sound Vibration 292:148–178

Kato M, Uematsu H, Kashio M, Hirahara T (2003) The effect of head motion on the accuracy of sound localization. Acoust Sci Tech 24:315–317

Katz BFG (2001) Boundary element method calculation of individual head–related transfer function. I. Rigid model calculation. J Acoust Soc Am 110:2440–2448

Kistler DJ, Wightman FL (1992) A model of head–related transfer functions based on principal components analysis and minimum–phase reconstruction. J Acoust Soc Am 91:1637–1647

Knudsen EI, Konishi M (1979) Mechanisms of sound localization in the barn owl (Tyto alba). J Comp Physiol 133:13–21
Kulkarni A, Colburn HS (1998) Role of spectral detail in sound–source localization. Nature 396:747–749
Kulkarni A, Isabelle SK, Colburn HS (1999) Sensitivity of human subjects to head-related transfer-function phase spectra. J Acoust Soc Am 105:2821–2840
Langendijk EHA, Bronkhorst AW (2002) Contribution of spectral cues to human sound localization. J Acoust Soc Am 112:1583–1596
Lord Rayleigh (1877) Acoustical observations. Phil Mag 3, 6th series: 456–464
Lord Rayleigh (1907) On our perception of sound direction. Phil Mag 13, 6th series: 214–232
Macpherson EA, Sabin AT (2013) Vertical–plane sound localization with distorted spectral cues. Hear Res 306:76–92
Mehrgardt S, Mellert V (1977) Transformation characteristics of the external human ear. J Acoust Soc Am 61:1567–1576
Middlebrooks JC (1992) Narrow–band sound localization related to external ear acoustics. J Acoust Soc Am 92:2607–2624
Middlebrooks JC (1999) Individual differences in external–ear transfer functions reduced by scaling in frequency. J Acoust Soc Am 106:1480–1492
Middlebrooks JC, Green DM (1992) Observations on a principal components analysis of head–related transfer functions. J Acoust Soc Am 92:597–599
Mokhtari P, Takemoto H, Nishimura R, Kato H (2007) Comparison of simulated and measured HRTFs: FDTD simulation using MRI head data. 123rd Audio Engineering Society Convention, New York, Preprint 7240: 1–12
Møller H (1992) Fundamentals of binaural technology. Appl Acoust 36:171–218
Morimoto M, Ando Y (1980) On the simulation of sound localization. J Acoust Soc Jpn (E) 1:167–174
Morimoto M, Saito A (1977) On sound localization in the median plane – effects of frequency range and intensity of stimuli. Technical report of technical committee of psychological and physiological acoustics, Acoust Soc Jpn H-40-1 (in Japanese)
Morimoto M, Joren N, Ando Y, Maekawa Z (1976) On the head-related transfer function. Technical report of technical committee of psychological and physiological acoustics. Acoust Soc Jpn H-31-1 (in Japanese)
Musicant AD, Butler RA (1984) The influence of pinnae-based spectral cues on sound localization. J Acoust Soc Am 75:1195–1200
Nojima R, Morimoto M, Sato H, Sato H (2013) Do spontaneous head movements occur during sound localization? J Acoust Sci & Tech 34:292–295
Otani M, Ise S (2003) A fast calculation method of the head–related transfer functions for multiple source points based on the boundary element method. Acoust Sci Tech 24:259–266
Perrett S, Noble W (1997) The effect of head rotations on vertical plane sound localization. J Acoust Soc Am 104:2325–2332
Raykar VC, Duraiswami R, Yegnanarayana B (2005) Extracting the frequencies of the pinna spectral notches in measured head related impulse responses. J Acoust Soc Am 118:364–374
Reiss LAJ, Young ED (2005) Spectral edge sensitivity in neural circuits of the dorsal cochlear nucleus. J Neuroscience 25:3680–3691
Sayers BM (1964) Acoustic-image lateralization judgement with binaural tones. J Acoust Soc Am 36:923–926
Seeber BU, Fastl H (2003) Subjective selection of non–individual head–related transfer functions. In: Proceedings of the 2003 international conference on auditory display, Boston
Shaw EAG (1997) Acoustical features of the human external ear. In: Gilkey RH, Anderson TR (eds) Binaural and spatial hearing in real and virtual environments. Erlbaum, Mahwah, pp 25–47
Shaw EAG, Teranishi R (1968) Sound pressure generated in an external-ear replica and real human ears by a nearby point source. J Acoust Soc Am 44:240–249

Takemoto H, Mokhtari P, Kato H, Nishimura R, Iida K (2012) Mechanism for generating peaks and notches of head-related transfer functions in the median plane. J Acoust Soc Am 132:3832–3841
Thurlow WR, Runge PS (1967) Effect of induced head movements on localization of direction of sounds. J Acoust Soc Am 42:480–487
Toole FE, Sayers BM (1965) Lateralization judgements and the nature of binaural acoustic images. J Acoust Soc Am 37:319–324
Wiener FM, Ross DA (1946) The pressure distribution in the auditory canal in a progressive sound field. J Acoust Soc Am 18:401–408
Wightman FL, Kistler DJ (1989a) Headphone simulation of free-field listening. I: Stimulus synthesis. J Acoust Soc Am 85:858–867
Wightman FL, Kistler DJ (1989b) Headphone simulation of free-field listening. II: Psychophysical validation. J Acoust Soc Am 85:868–878
Xiao T, Liua QH (2003) Finite difference computation of head–related transfer function for human hearing. J Acoust Soc Am 113:2434–2441
Zotkin DN, Hwang J, Duraiswami R, Davis LS (2003) HRTF Personalization using anthropometric measurements. In: IEEE workshop on applications of signal processing to audio and acoustics
Zotkin DN, Duraiswami R, Grassi E, Gumerov NA (2006) Fast head–related transfer function measurement via reciprocity. J Acoust Soc Am 120:2202–2215

Chapter 2
HRTF and Sound Localization in the Horizontal Plane

Abstract This chapter describes the physical aspect of the HRTFs in the horizontal plane and sound image localization by reproduction of the HRTFs in the horizontal plane. Then, the cues for lateral localization, which are included in HRTFs, are discussed.

2.1 HRTF in the Horizontal Plane

Figure 2.1 shows the amplitude spectra of the HRTFs for the sound source in the horizontal plane from the front (azimuth angle of 0°) to the rear (azimuth angle of 180°) in 30° steps. The solid red lines indicate the HRTFs of the right ear (ipsilateral to a sound source), and the blue dotted lines indicate those of the left ear (contralateral to a sound source). The unit of the ordinate is indicated as 10 dB. Each HRTF is drawn with a 40-dB shift. The broken line indicates the 0-dB line for each HRTF. Figure 2.1 shows the following:

1. The notch and peak are distinct in the HRTFs of the ipsilateral ear. Some peaks exceed 10 dB, and some notches exceed −20 dB. The frequencies of the notches are shifted higher as the sound source moves from the front to the rear, while the frequencies of the peaks are approximately constant, independent of the azimuth angle of a sound source.
2. For a sound source located in the lateral direction, the spectrum of the HRTF of the ear contralateral to the sound source is relatively flat because the effect of the pinnae is slight.
3. The interaural level difference reaches a maximum for a sound source located in the lateral direction, and exceeds 30 dB at a particular frequency.
4. The HRTFs of the left and right ears are not identical, even for a sound source located in the front or rear directions. This is due to asymmetry of the head and pinnae.

Thus, the amplitude spectrum of the HRTF and the interaural level difference vary with the azimuth angle of a sound source.

K. Iida, *Head-Related Transfer Function and Acoustic Virtual Reality*,
https://doi.org/10.1007/978-981-13-9745-5_2

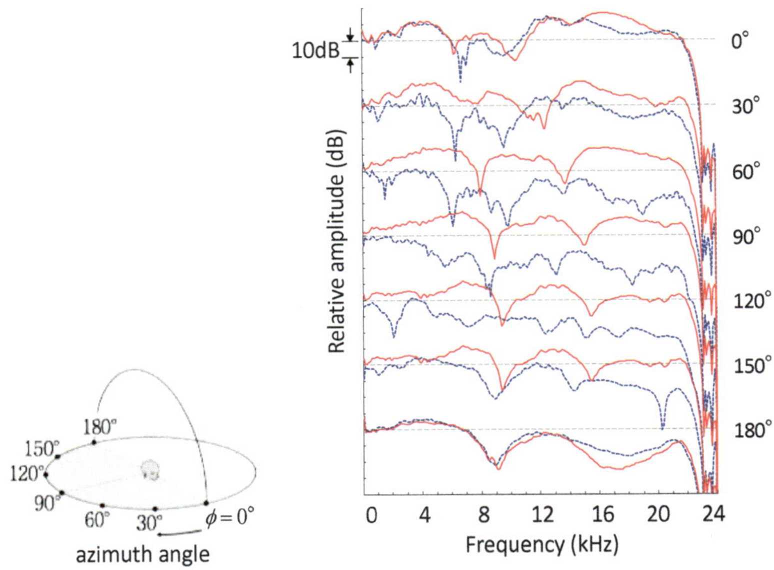

Fig. 2.1 Examples of amplitudes of HRTFs for the sound source in the horizontal plane (0°–180°). The solid red lines and the blue dotted lines indicate the HRTFs of the right and left ears, respectively

2.2 Sound Localization in the Horizontal Plane

2.2.1 Localization Using the listener's Own HRTFs

The information of the incident angle of sound exists in the HRTF. Therefore, in principle, accurate virtual sound image localization must be achieved when the listener's own HRTFs are reproduced at the entrances of the ear canals. Here, I introduce the results of two kinds of sound localization tests reproducing the HRTFs in the horizontal plane: (1) through headphones and (2) by a transaural system using two loudspeakers.

1. *Reproduction of HRTFs through headphones*

In the 1980s, it was demonstrated that the direction of a sound image is accurately reproduced by reproducing the listener's own HRTF at the ear drum (Wightman and Kistler 1989). However, this method is dangerous for practical application because a probe-microphone must be placed in close proximity to the listener's eardrum. Moreover, the experimental system based on this method is difficult to use, even for research in the laboratory. Later, an algorithm, in which the HRTFs at the entrances of the open ear canals (usual listening condition) can be reproduced

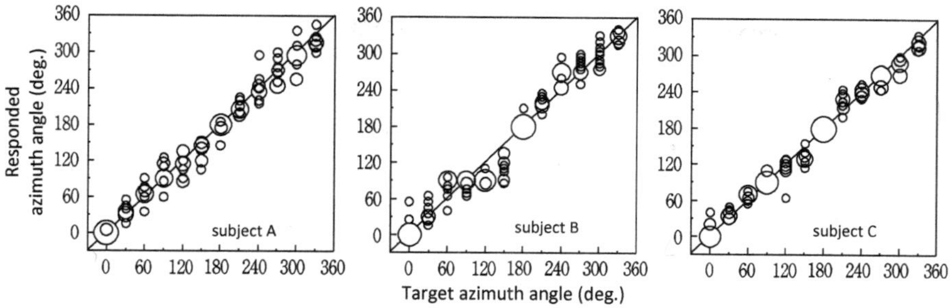

Fig. 2.2 Sound image localization in the horizontal plane by reproducing the listener's own HRTFs through headphones. The radius of each circle is proportional to the number of responses with a resolution of 5°

using the HRTFs measured at the entrances of the blocked ear canals (a condition in which HRTFs are easy to measure) was developed (Møller et al. 1995a, b) (see Sect. 12.1 for details). This algorithm enabled the easy and accurate reproduction of HRTFs through headphones.

Figure 2.2 shows the results of the localization tests obtained using this algorithm. The abscissa indicates the azimuth angle of the HRTFs in the horizontal plane (hereinafter referred to as the target azimuth angle) from 0° to 330° in 30° steps. The ordinate indicates the responded azimuth angle by subjects. The subjects were three males (A, B, and C) in their twenties. The sound source was a wide-band white noise from 200 Hz to 17 kHz. The responses were distributed along a diagonal line. This indicates that the listeners localized a sound image around the target azimuth angle when their own HRTFs were reproduced.

2. *Reproduction of HRTFs by a transaural system using two loudspeakers*

Figure 2.3 shows the results of the localization tests (Morimoto and Ando 1980), in which the subject's own HRTFs in the horizontal plane (0°–180°, in 15° steps) were reproduced at the entrances of the subject's ear canals by the transaural system,[1] which consists of a digital filter matrix and two loudspeakers (see Chap. 12.2 for details). The tests were performed in an anechoic chamber. The subjects were three males (L, M, and S) in their twenties. The sound source was wide-band white noise from 300 Hz to 13.6 kHz.

The responses of each subject were distributed along a diagonal line, although the responses of subject L tended to be shifted slightly laterally, as compared to the target azimuth angle.

[1]Note that the heads of the subjects were fixed tightly using a jig in the tests. There exist some requirements that must be satisfied in order to ensure accurate reproduction of the direction of a sound image by a transaural system (see Sect. 12.2 for details). Thus, at present, the transaural system is difficult to use as an experimental tool.

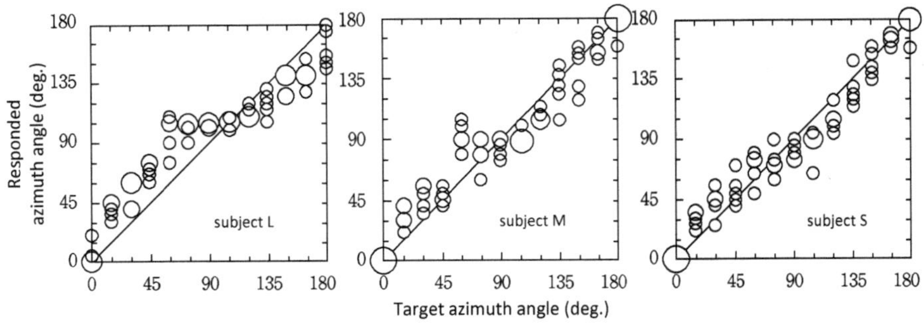

Fig. 2.3 Sound image localization in the horizontal plane by reproducing the listener's own HRTFs using the transaural system (Morimoto and Ando 1980). The area of each circle is proportional to the number of responses with a resolution of 5°

Thus, the direction of a sound image in the horizontal plane can be reproduced by reproducing the listener's own HRTFs using either the headphones or the transaural system.

2.2.2 Localization Using Others' HRTFs

It is not easy to obtain the listener's own HRTFs (as discussed in detail in Chap. 4). Therefore, there is great interest in sound localization using others' HRTFs from the standpoint of the practical use of acoustic VR.

Figure 2.4 shows the results of the localization tests, in which others' HRTFs in the horizontal plane (0°–180°, in 15° steps) were reproduced by the transaural system (Morimoto and Ando 1980). The experimental conditions are the same as for Fig. 2.3, except for the reproduced HRTFs. Panels (a) and (b) show the responses of subject L to the HRTFs of subjects M and S, respectively, and panels (c) and (d) show the responses of subject S to the HRTFs of subjects L and M, respectively. The designations of the subjects (i.e., subjects S, M, and L) indicate the relative size of the subject's pinnae (i.e., small, medium, and large, respectively).

Subject L can discriminate the left-right direction of a sound image for the HRTFs of subjects M and S. However, Subject L localized a sound image at the rear for target azimuth angles of 15°–75°. Moreover, for the HRTFs of subject S, subject L sometimes localized a sound image at 0°, and, at other times localized a sound image at 180° for a target azimuth angle of 0°.

On the other hand, subject S localized a sound image around the target azimuth angle for the HRTFs of subjects L and M, although subject S tended to localize a sound image at the rear for target azimuth angles of 15° and 30° for the HRTFs of subject L. Although not established theoretically, the HRTFs of large pinnae tend to provide good localization performance to listeners who have small pinnae (see Sect. 3.8 for details).

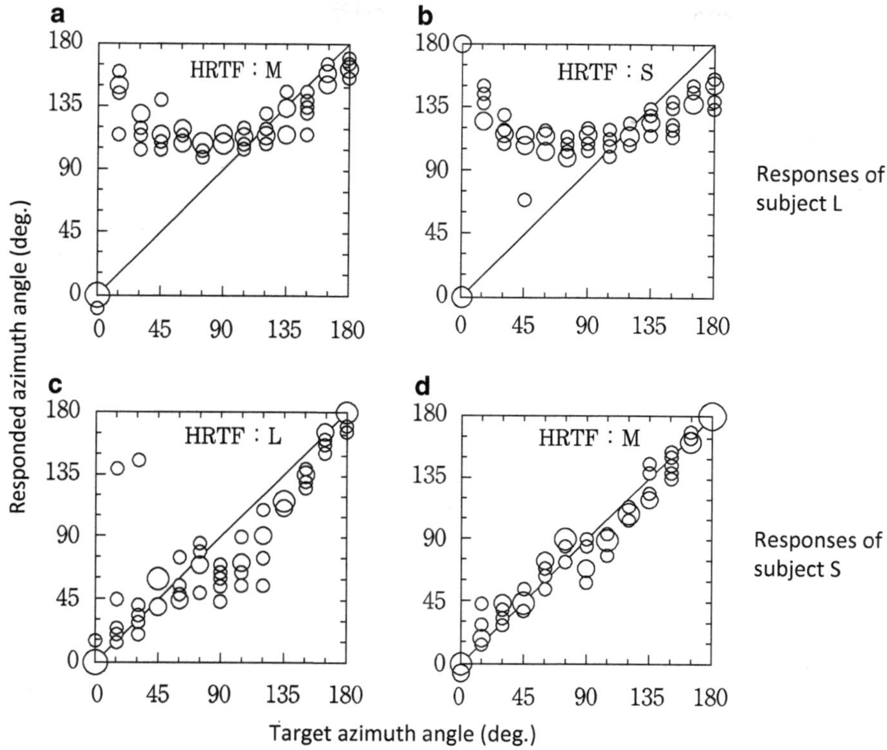

Fig. 2.4 Sound image localization in the horizontal plane by reproducing others' HRTFs using the transaural system. Panels (**a**) and (**b**) show the responses of subject L to the HRTFs of subjects M and S, respectively, and panels (**c**) and (**d**) show the responses of subject S to the HRTFs of subjects L and M, respectively. (Morimoto and Ando 1980)

Thus, reproducing others' HRTFs provides accurate sound image reproduction for the left-right direction, but often causes front-back error.

2.3 Cues for Lateral Localization

As mentioned in Sect. 2.2, the direction of a sound image in the horizontal plane is accurately reproduced by reproducing the listener's own HRTFs. In addition, left-right discrimination can be achieved even when others' HRTFs are used. Here, I will advance the discussion on what parts of the HRTF are cues for lateral localization.

The human ears are situated on either side of the head. As such, differences in arrival time and sound pressure between the left and right ears are generated when the sound comes from the side. Around 1900, it was reported that the cues for left-right localization are the interaural time difference (ITD) and the interaural level difference (ILD) (Lord Rayleigh 1877, 1907).

2.3.1 Interaural Time Difference (ITD)

Figure 2.5 shows the ITDs for the sound source in the horizontal plane obtained for 33 adults (see Sect. 10.1 for the detailed calculation method). The absolute value of the ITD is approximately 0 for the front and rear directions and ranges from 600 to 800 μs for the lateral direction.

Regarding an incident sound wave as a plane wave and a head as a perfect sphere, the relationship between the incident azimuth angle and the ITD can be represented by Eq. (2.1) (Blauert 1997a).

$$\Delta S = c \times ITD = \frac{D}{2}(\phi + \sin\phi) \tag{2.1}$$

where ΔS is the path difference between both ears, c is the speed of sound, and ITD is the interaural time difference. Moreover, D and ϕ denote the distance between both ears (diameter of the sphere) and the incident azimuth in radians, respectively.

The results of the experiments on the relationship between the ITD and the lateral localization, in which no HRTFs were reproduced, but only the ITD was controlled through earphones, showed that the subjects localized a sound image to the front direction for the ITD of 0 ms and to the lateral direction for the ITD of ± 1 ms (Blauert 1997b).

The ITD is not merely a numerical model. It has been reported that the function of calculating the time difference between input signals to the left and right ears exists in the medial nucleus of the superior olive in the human auditory system. The superior olive is located on the pathway from the inner ear to the primary auditory cortex.

Figure 2.6 shows the coincidence model of the ITD reported by Jeffress-Colburn. In the figure, Δ, $\times$, and $\int$ indicate the time delay, a multiplier, and an integrator.

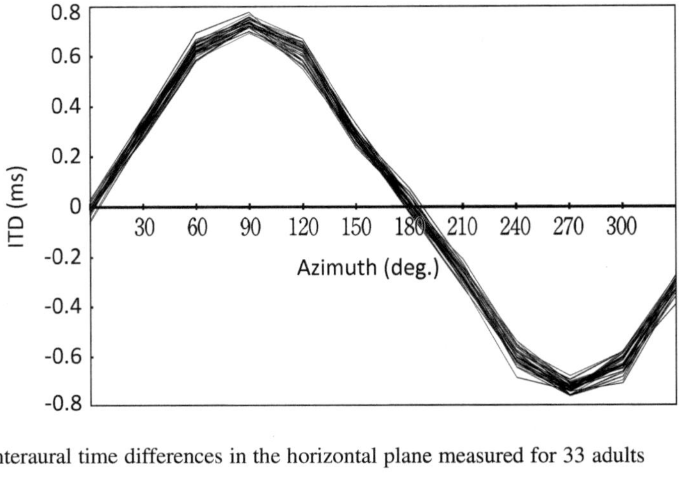

Fig. 2.5 Interaural time differences in the horizontal plane measured for 33 adults

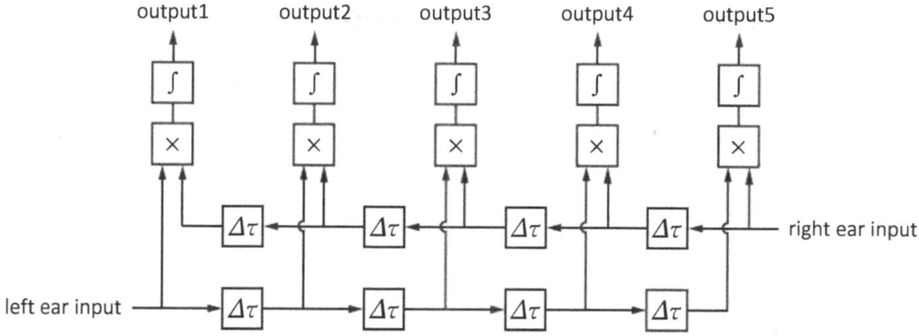

Fig. 2.6 Coincidence model of ITD proposed by Jeffress-Colburn

The integrator processes the smoothing of the signal in the time domain. In the model, the location, in which the maximum is obtained, varies with the ITD. For instance, in the case that the sound comes from the left side of the head, the maximum value is obtained at the right edge of the delay line (output 5).

The frequency range, in which the fine waveform of the ear-input signals is used in the calculation of the ITD, is limited to below 1600 Hz. This is caused by the absolute refractory period of the firing of the neural pulse at the cochlea. Above 1600 Hz, the human auditory system calculates the ITD from the envelop of the ear-input signals.

2.3.2 Interaural Level Difference (ILD)

The interaural level difference also varies with the incident azimuth angle of a sound wave. Figure 2.7 shows the ILD in the horizontal plane for the same 33 adults as Fig. 2.5. The absolute value of the ILD is approximately 0 at the front and rear directions, and reaches the maximum at the lateral direction. The absolute value of the ILD becomes large as the frequency becomes high for the same azimuth angle. The average ILD across the 33 subjects for the azimuth angle of 90° is 4.3 dB for the 1/3 octave band of the center frequency of 250 Hz, and those for 1 kHz and 4 kHz are 10.1 dB and 18.4 dB, respectively. This difference is caused by the frequency dependence of the diffraction of a sound wave.

The results of the experiments on the relationship between the ILD and the lateral localization for a pure tone of 600 Hz and a wide-band noise, in which no HRTFs were reproduced, but only the ILD was controlled, showed that the subjects localized a sound image to the front direction with the ILD of 0 dB, and to the lateral direction with the ILD of ±12 dB for both the pure tone of 600 Hz and the wide-band noise (Blauert 1997c).

It has been reported that the function for calculating the level difference between the input signals to the left and right ears exists in the lateral nucleus of the superior

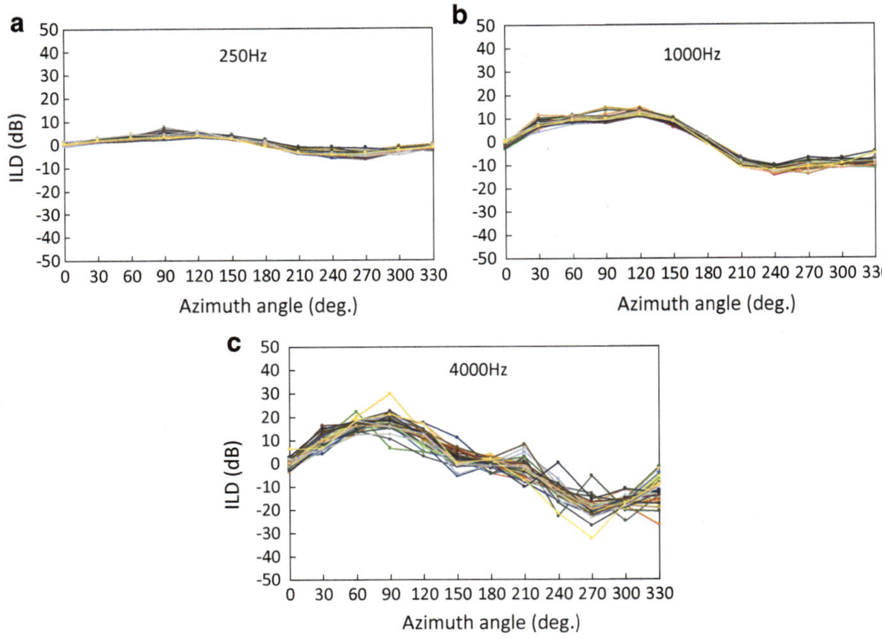

Fig. 2.7 Interaural level differences in the horizontal plane measured for 33 adults for the 1/3 octave bands of (**a**) 250 Hz, (**b**) 1 kHz, and (**c**) 4 kHz

olive in the human auditory system. The ILD acts as a cue for the lateral localization over the entire audible frequency range.

2.4 Cone of Confusion

Assuming the head is spherical, the interaural differences are identical at any points on the circle of the vertical section (sagittal plane) of the cone shown in Fig. 2.8. Therefore, the information of the interaural difference only indicates in which sagittal plane a sound source is located. The interaural difference does not provide the vertical angle of the sound source in the sagittal plane. This is called the cone of confusion.

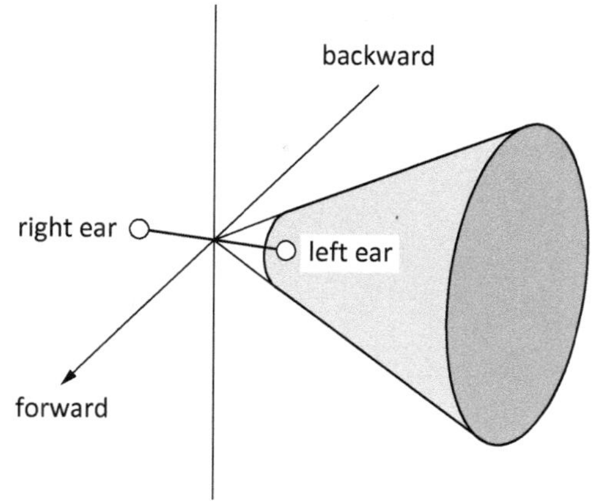

Fig. 2.8 Cone of confusion

2.5 Summing Sound Image Localization Using Multiple Sound Sources

One of the technological applications of the ILD is the stereo system. The stereo system controls the lateral direction of a sound image by the ILD. The ILD is controlled by changing the balance of the output level of the left and right loudspeakers. As such, the localization of a single sound image by reproducing an identical sound source through multiple loudspeakers is called summing localization. In summing localization, the position of a sound image can be controlled at an arbitrary point on the line segment connecting two loudspeakers, as shown in Fig. 2.9.

Then, is a sound image perceived at the lateral direction when two loudspeakers are placed at the front and rear directions of a listener? Figure 2.10 shows the results of the sound localization tests, in which two loudspeakers were placed at azimuth angles of 240° and 300° (Theile and Plenge 1977). The output sound pressure level difference between two loudspeakers was varied within ±18 dB. In the case of a level difference of more than 6 dB, a sound image was localized at the direction of the louder loudspeaker. In the case of a difference of less than 6 dB, the direction of the sound image changed abruptly. In other words, the stable sound image was not perceived at the lateral direction. This implies that it is impossible to control the front-back direction of a sound image by the difference in output sound pressure of two loudspeakers. As described in Chap. 3, the cue for front-back perception is not the ILD. The loudspeakers placed at the lateral direction are needed in order to control a sound image over the entire circumference in the horizontal plane by controlling the difference in output sound pressure level of the loudspeakers (panning control).

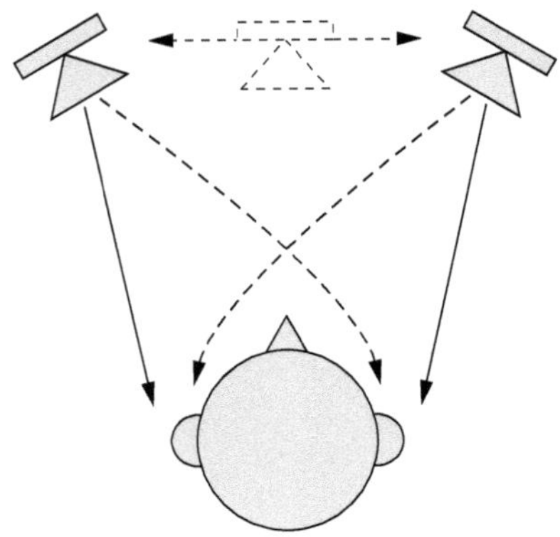

Fig. 2.9 Summing localization using the output sound pressure level difference between two loudspeakers

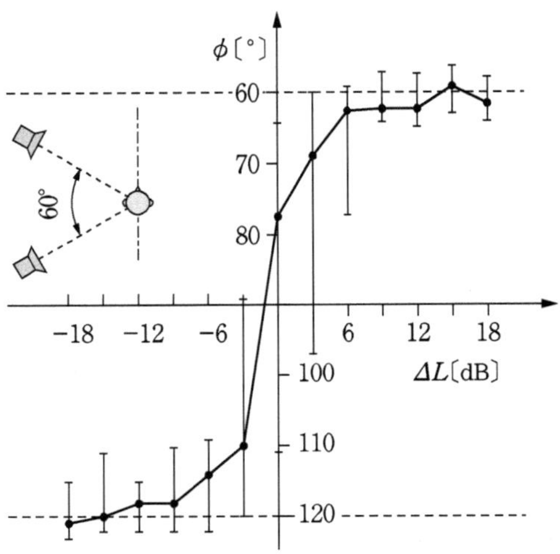

Fig. 2.10 Relationship between the output sound pressure level difference between two loudspeakers placed at azimuth angles of 240° and 300° and the direction of a sound image (Theile and Plenge 1977)

References

Blauert J (1997a) Spatial hearing revised edition. The MIT Press, Cambridge, p 76

Blauert J (1997b) Spatial hearing revised edition. The MIT Press, Cambridge, p 144

Blauert J (1997c) Spatial hearing revised edition. The MIT Press, Cambridge, p 158

Møller H, Hammershøi D, Jensen CB, Sørensen MF (1995a) Transfer characteristics of headphones measured on human ears. J Audio Eng Soc 43:203–217

Møller H, Sørensen MF, Hammershøi D, Jensen CB (1995b) Head-related transfer functions of human subjects. J Audio Eng Soc 43:300–321

Morimoto M, Ando Y (1980) On the simulation of sound localization. J Acoust Soc Jpn (E) 1:167–174

Lord Rayleigh (1877) Acoustical observations. Phil. Mag. 3, 6th series: 456–464

Lord Rayleigh (1907) On our perception of sound direction. Phil. Mag. 13, 6th series: 214–232

Theile G, Plenge G (1977) Localization of lateral phantom sources. J Audio Eng Soc 25:196–200

Wightman FL, Kistler DJ (1989) Headphone simulation of free-field listening. II: Psychophysical validation. J Acoust Soc Am 85:868–878

Chapter 3
HRTF and Sound Localization in the Median Plane

Abstract This chapter describes the physical aspects of HRTFs in the median plane and sound image localization by reproducing these HRTFs. Cues for vertical localization, which are included in HRTFs, are also discussed.

3.1 HRTFs in the Median Plane

Figure 3.1 shows the amplitude spectra of the HRTFs for a sound source in the median plane from the front (vertical angle of 0°) to the rear (vertical angle of 180°) in 30° steps. The solid red lines indicate the HRTFs for the right ear, and the blue dotted lines indicate those for the left ear. A difference of 10 dB is indicated on the ordinate. Each HRTF is drawn with a 40-dB shift. The broken line indicates the 0-dB level for each HRTF. Figure 3.1 shows the following:

1) The frequency of the peak around 4 kHz is approximately constant, independent of the vertical angle of the sound source.
2) The frequencies of the notches shift higher as the sound source moves from the front of the subject to above the subject, and reach a maximum at a vertical angle of 120°.
3) The notches are deep for a sound source near the horizontal plane and are shallow for a sound source above the horizontal plane.
4) The HRTFs for the left and the right ears are not identical, even for a sound source located in the median plane. This is due to the asymmetry of the head and pinnae shape.

Thus, the amplitude spectrum of a HRTF varies with the vertical angle of the sound source.

K. Iida, *Head-Related Transfer Function and Acoustic Virtual Reality*,
https://doi.org/10.1007/978-981-13-9745-5_3

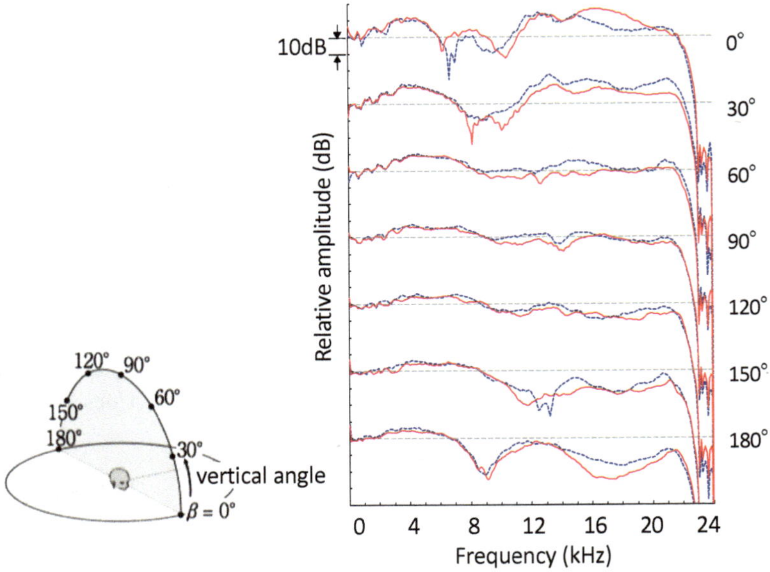

Fig. 3.1 Amplitudes of HRTFs for sound source in upper median plane (0°–180°). The solid red lines and blue dotted lines indicate HRTFs for the right and left ears, respectively

3.2 Sound Localization in the Median Plane

3.2.1 Localization Using listener's Own HRTFs

1. *Reproduction of HRTFs through headphones*

Figure 3.2 shows the results of localization tests in the upper median plane (0°–180°, in 30° steps), in which the listener's own HRTFs were reproduced through headphones using the method described in Sect. 2.2. The subjects were three males and one female, all in their twenties. The sound source was wide-band white noise from 200 Hz to 17 kHz.

Most of the responses are distributed along the solid diagonal line in the figure, whereas the responses for subjects MTZ and YMM are distributed as an inverted s-shaped curve above this diagonal line, which appeared in the responses to the actual sound sources (see Sect. A1.2). These results indicate that the listeners localized a sound image around the target vertical angle when their own HRTFs were reproduced.

2. *Reproduction of HRTFs by a transaural system using two loudspeakers*

Figure 3.3 shows the results of localization tests (Morimoto and Ando 1980) in which the subject's own HRTFs in the upper median plane (0°–180°, in 15° steps) were reproduced at the entrances of the subject's ear canals by a transaural system.

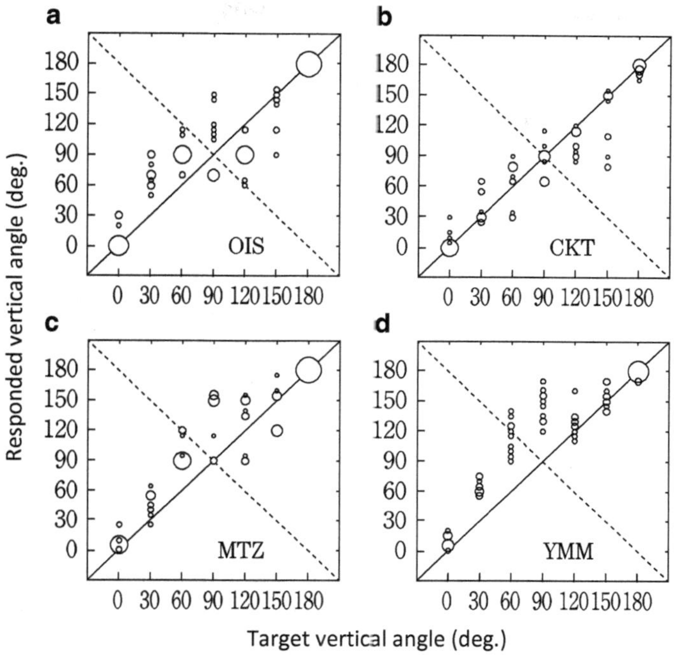

Fig. 3.2 Sound image localization in the upper median plane by reproducing listener's own HRTFs through headphones. The radius of each circle is proportional to the number of responses with a resolution of 5°. (**a**) subject OIS, (**b**) subject CKT, (**c**) subject MTZ, and (**d**) subject YMM

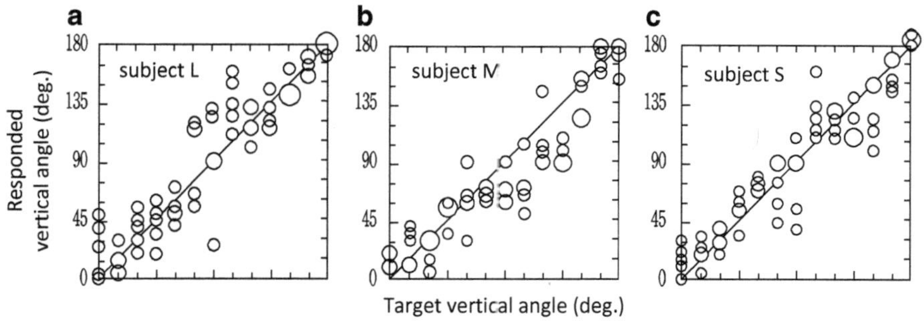

Fig. 3.3 Sound image localization in the upper median plane by reproducing listener's own HRTFs using transaural system. (**a**) subject L, (**b**) subject M, and (**c**) subject S. (Morimoto and Ando 1980)

The subjects were three males (L, M, and S) in their twenties. The sound source was wide-band white noise from 300 Hz to 13.6 kHz.

The responses for each subject are distributed along a diagonal line, although the variances of the responses were larger than those in the horizontal plane. This tendency is observed in localization tests using actual sound sources.

Thus, the direction of a sound image in the median plane can be reproduced by reproducing the listener's own HRTFs using either the headphones or the transaural system.

3.2.2 *Localization Using Others' HRTFs*

1. *Reproduction of HRTFs through headphones*

Figure 3.4 shows the results of localization tests in which others' HRTFs in the upper median plane (0°–180°, in 30° steps) were reproduced through headphones. The subjects are the same as for Fig. 3.2. The HRTFs of a male other than the subject were reproduced. Subject OIS localized a sound image around the target vertical angle. However, the other three subjects localized a sound image at 120°–180° for a target vertical angle of 0°. Moreover, the responses for subject MTZ are widely distributed from forward to rearward for a target vertical angle of 180°. Thus, reproducing other's HRTFs often causes front-back errors.

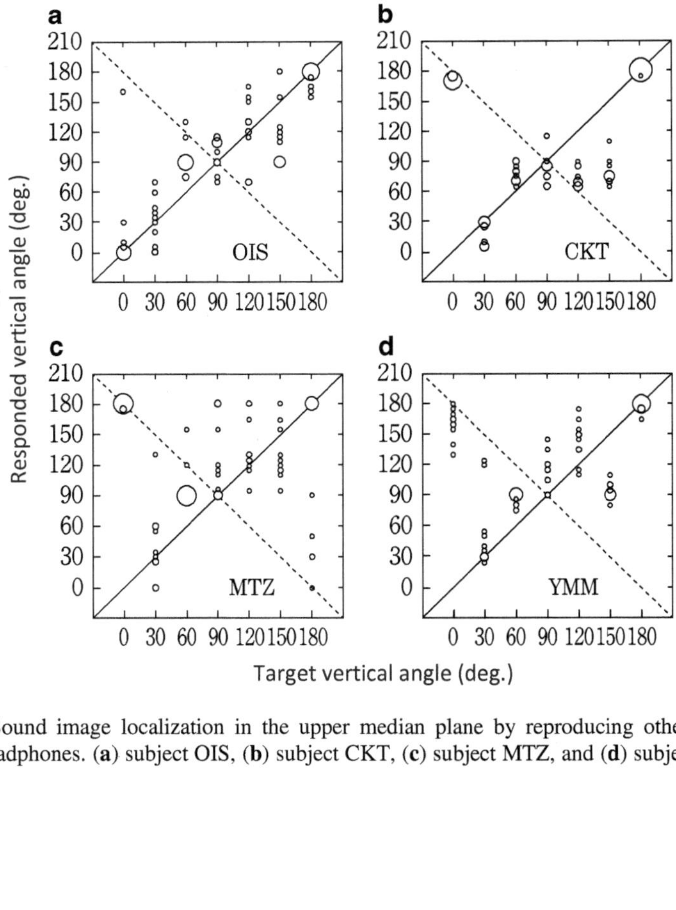

Fig. 3.4 Sound image localization in the upper median plane by reproducing other's HRTFs through headphones. (**a**) subject OIS, (**b**) subject CKT, (**c**) subject MTZ, and (**d**) subject YMM

2. *Reproduction of HRTFs by a transaural system using two loudspeakers*

Figure 3.5 shows the results of localization tests in which others' HRTFs in the upper median plane (0°–180°, in 15° steps) were reproduced by a transaural system (Morimoto and Ando 1980). The experimental conditions are the same as for Fig. 3.3, except for the reproduced HRTFs. Panels (a) and (b) show the responses for subject L to the HRTFs of subjects M and S, respectively, and panels (c) and (d) show the responses for subject S to the HRTFs of subjects L and M, respectively. The designations of the subjects (i.e., subjects S, M, and L) indicate the relative size of the subject's pinnae (i.e., small, medium, and large, respectively). Subject L never localized a sound image at the front for the HRTFs of subjects M and S. Most of the responses for subject L were distributed from 120° to 180°. The responses for subject S shifted upward for the HRTFs of subject L for target vertical angles of 0°–45°. The responses for subject S were widely distributed for the HRTFs of subject M for target vertical angles of 45°–120°.

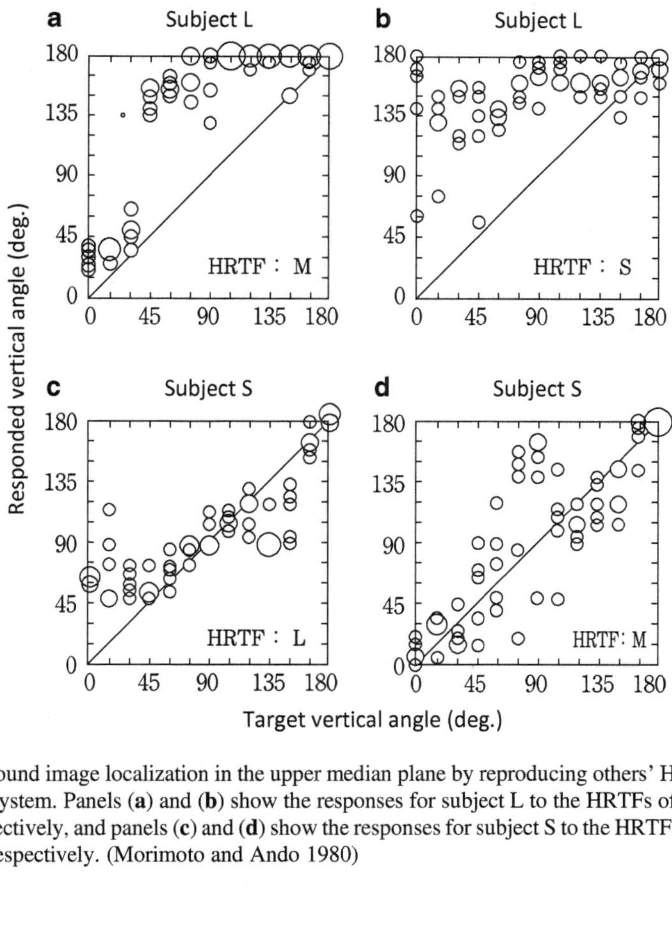

Fig. 3.5 Sound image localization in the upper median plane by reproducing others' HRTFs using transaural system. Panels (**a**) and (**b**) show the responses for subject L to the HRTFs of subjects M and S, respectively, and panels (**c**) and (**d**) show the responses for subject S to the HRTFs of subjects L and M, respectively. (Morimoto and Ando 1980)

3.2.3 *Three Major Errors Regarding Sound Localization in the Median Plane*

Subjects often perceive a sound image in a direction other than the target direction when other's HRTFs are reproduced. The errors fall roughly into the following three categories (Fig. 3.6): (1) front-back confusion, (2) rising of a sound image, and (3) inside-of-head localization (lateralization).

1. **Front-back confusion**

Front-back confusion is a phenomenon in which the target direction and the perceived direction are reversed back to front.

Since the frequencies of the prominent notches (the cues for the front-back direction are described in Sect. 3.3) of other's HRTFs do not often coincide with those of the listener's HRTFs, listeners obtain inadequate information for sound image localization.

Listeners often perceive a sound image to the front for a rear target direction when other's HRTFs are reproduced by a transaural system in which the loudspeakers are visible. This might be explained by visual information affecting the direction of a sound image when the aural information is ambiguous. On the other hand, listeners often perceive a sound image at the rear for a front target direction when other's HRTFs are reproduced through headphones, in which case the loudspeakers are invisible.

2. **Rising of a sound image**

A sound image often rises from the horizontal plane, even though the target direction is on the horizontal plane, when other's HRTFs are reproduced. The reason why sound images rise, but do not fall, is unclear.

3. **Inside-of-head localization (lateralization)**

When diotic sound signals without HRTFs are reproduced at the entrances of the ear canals, listeners perceive a sound image inside their heads. The human auditory system cannot detect vertical angle information in ear-input signals, which does not include HRTFs, and, as a result, the listener perceives a sound image inside his/her head.

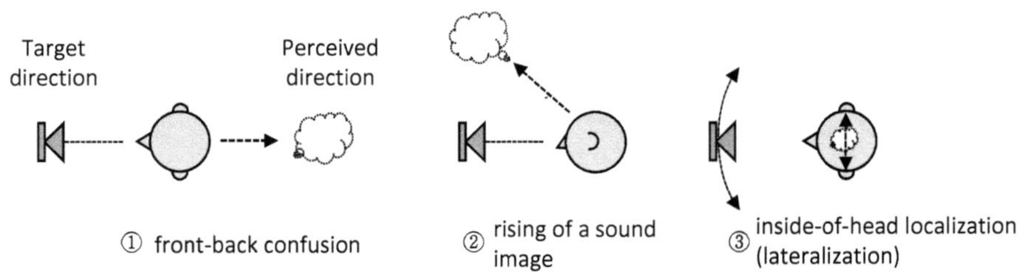

Fig. 3.6 Three major localization errors when other's HRTFs are reproduced

3.3 Cues for Vertical Localization

3.3.1 *Overview of Spectral Cues*

What are the cues for vertical localization? After the 1970s, a number of studies examined cues for the perception of vertical direction. Consequently, it was found that cues exist in the amplitude spectrum of the HRTF. These are referred to as spectral cues. Moreover, studies to find the specific important part of the amplitude spectrum that acts as a spectral cue have been performed.

These studies revealed that spectral cues exist in the frequency range above 5 kHz. Figure 3.7 shows the effect of the frequency range of the sound source on the accuracy of median plane localization (Morimoto and Saito 1977). For wide-band white noise (a) and noise with a low-pass cut-off frequency of 9.6 kHz (b), the responses are distributed along a diagonal line. For noise with a low-pass cut-off frequency of 4.8 kHz (c), however, the subjects never localized a sound image in the upper direction. For a cut-off frequency of 2.4 kHz (d), front-back errors were observed. On the other hand, for high-pass-filtered noise (e) through (g), the variance of the distribution of the responses tended to be large as the cut-off frequency increased. These results suggest that the frequency components between 5 kHz and 10 kHz are necessary for accurate median plane localization.

Moreover, it has been reported that the frequency components above 16 kHz and below 3.8 kHz do not affect the accuracy of median plane localization (Hebrank and Wright 1974). However, for the sagittal plane far from the median plane, cues for vertical localization were reported to exist below 3 kHz (Algazi et al. 2001).

It is widely known that spectral notches and peaks above 5 kHz contribute to the perception of the vertical angle of a sound image (Hebrank and Wright 1974; Butler and Belendiuk 1977; Mehrgardt and Mellert 1977; Musicant and Butler 1984). The frequency of notches shifts higher as the sound source moves from the front of the subject to above the subject (Butler and Belendiuk 1977; Shaw and Teranishi 1968). The notches are generated in the pinnae (Musicant and Butler 1984; Gardner and Gardner 1973; Lopez–Poveda and Meddis 1996; Iida et al. 1998; Takemoto et al. 2012), and the frequency of the notches depends on the shape of the pinnae as well as the vertical angle (Raykar et al. 2005). Moreover, the outline of the amplitude spectrum has been found to be more important than its fine structure (Asano et al. 1990; Middlebrooks 1992, 1999; Kulkarni and Colburn 1998; Langendijk and Bronkhorst 2002; Macpherson and Sabin 2013). The difference in notch frequency due to the vertical angle has been reported to be detectable by the listener (Moore et al.1989).

3.3.2 *Details of Spectral Cues*

A parametric HRTF model, recomposed of all or some of the spectral notches and peaks extracted from a listener's own HRTF, taking the peak around 4 kHz as the lower-frequency limit, has been proposed (Iida et al. 2007). These notches and peaks are labeled

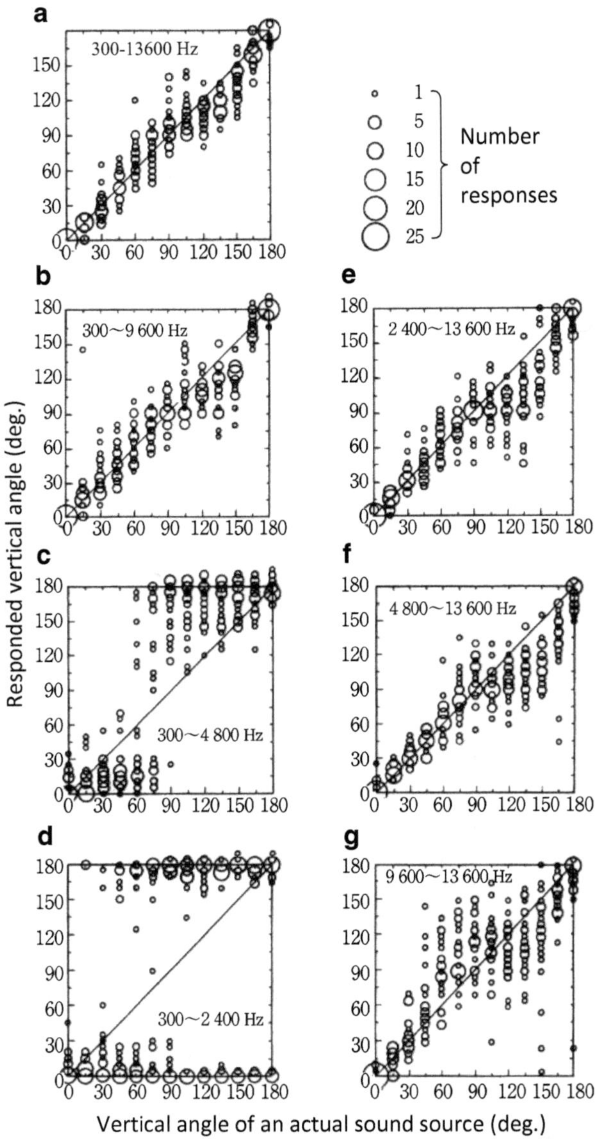

Fig. 3.7 Effects of frequency range of stimuli (white noise) on the median plane localization. (Morimoto and Saito 1977)

in order of frequency (e.g., N1, P1, N2, P2, N3, P3, and so on), and each is expressed parametrically in terms of frequency, level, and sharpness, as shown in Fig. 3.8.

It has been reported that there exist six prominent spectral peaks up to 20 kHz in the HRTFs in the median plane (Shaw 1997; Kahana and Nelson 2006). Figure 3.9 shows a schematic representation of the three lowest-frequency peaks (P1, P2, and P3) and the three lowest-frequency notches (N1, N2, and N3) (Takemoto et al. 2012). The frequencies of the peaks are approximately constant and are thus independent of the vertical angle, whereas the frequencies of the notches are highly dependent on the vertical angle. This vertical angle dependency of the notch

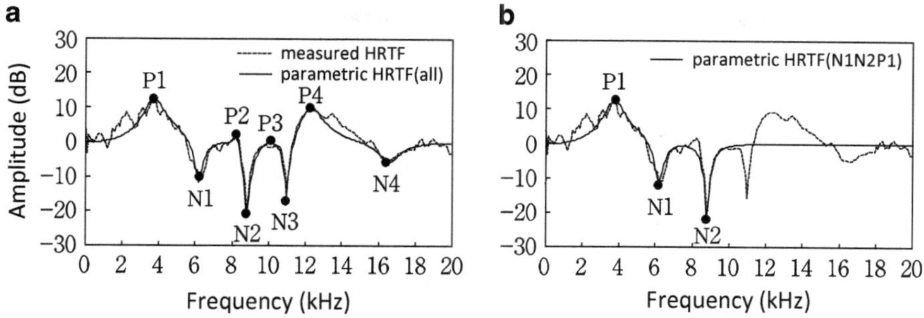

Fig. 3.8 Examples of parametric HRTFs. (**a**) Recomposed of all notches and peaks, (**b**) Recomposed of N1, N2, and P1

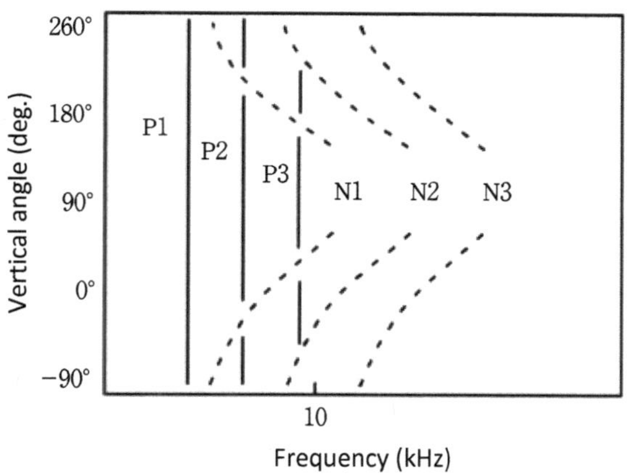

Fig. 3.9 Schematic diagram of notches and peaks in the median plane

frequency is thought to contribute as one of the important cues for vertical localization.

Localization tests and observations of the spectral peaks and notches of the HRTFs in the upper median plane infer the following.

1. **Minimum components of notches and peaks for median plane localization**

We first consider the minimum required number of notches and peaks for accurate median plane localization.

Figure 3.10(a) and (b) show the results of median plane localization tests with parametric HRTFs recomposed of all the notches and peaks extracted from the listener's own HRTFs. The results are seen to be similar to those using the measured HRTFs, as shown in Figs. 3.10(c) and (d) (Iida et al. 2007).

As shown in Fig. 3.11, localization tests were also carried out with parametric HRTFs recomposed of only some of the spectral notches and peaks extracted from the listener's own HRTFs. The results demonstrated that the two lowest-frequency notches (N1 and N2) and the lowest-frequency peak (P1) provided approximately

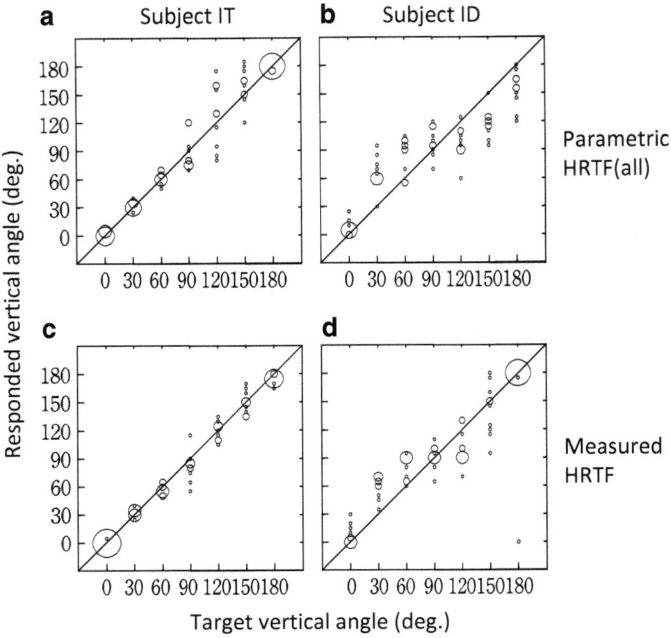

Fig. 3.10 Responses for parametric HRTFs using all notches and peaks and measured HRTFs. (**a**) and (**b**) parametric HRTFs recomposed of all the notches and peaks, (**c**) and (**d**) measured own HRTFs. (Iida et al. 2007)

the same localization performance as the listener's own HRTFs for the front and rear directions (Fig. 3.11(e) to (h)) (Iida et al. 2007).

For the upper directions, however, the localization performance of the parametric HRTF recomposed of N1, N2, and P1 for some of the subjects decreased compared with the subject's own HRTFs (Fig. 3.11(e) to (g)).

Median plane localization tests were then carried out using parametric HRTFs constructed using N1, N2, P1, and P2 (Iida and Ishii 2018). Figure 3.11(i) to (l) show that the localization performance for all subjects was improved at certain target vertical angles by adding P2 to N1N2P1. For subject MKI, the performance at 120° and 150° was improved. For subject OIS, the performance at 90° was improved. For subject OTK, the scatter in the responses observed at 60°, 90°, and 120° for N1N2P1 was not found for N1N2P1P2. The distribution of responses for N1N2P1P2 was approximately the same as that for the measured HRTFs. For subject YSD, the localization performance at 0° was improved. The difference in mean vertical localization errors between N1N2P1P2 and the measured HRTFs became less than 10° for all seven vertical target angles.

These results imply that N1N2P1P2 provides approximately the same vertical localization performance as the measured HRTFs at any of the seven target vertical angles in the upper median plane. In other words, the minimum set of notches and peaks for accurate median plane localization is N1, N2, P1, and P2.

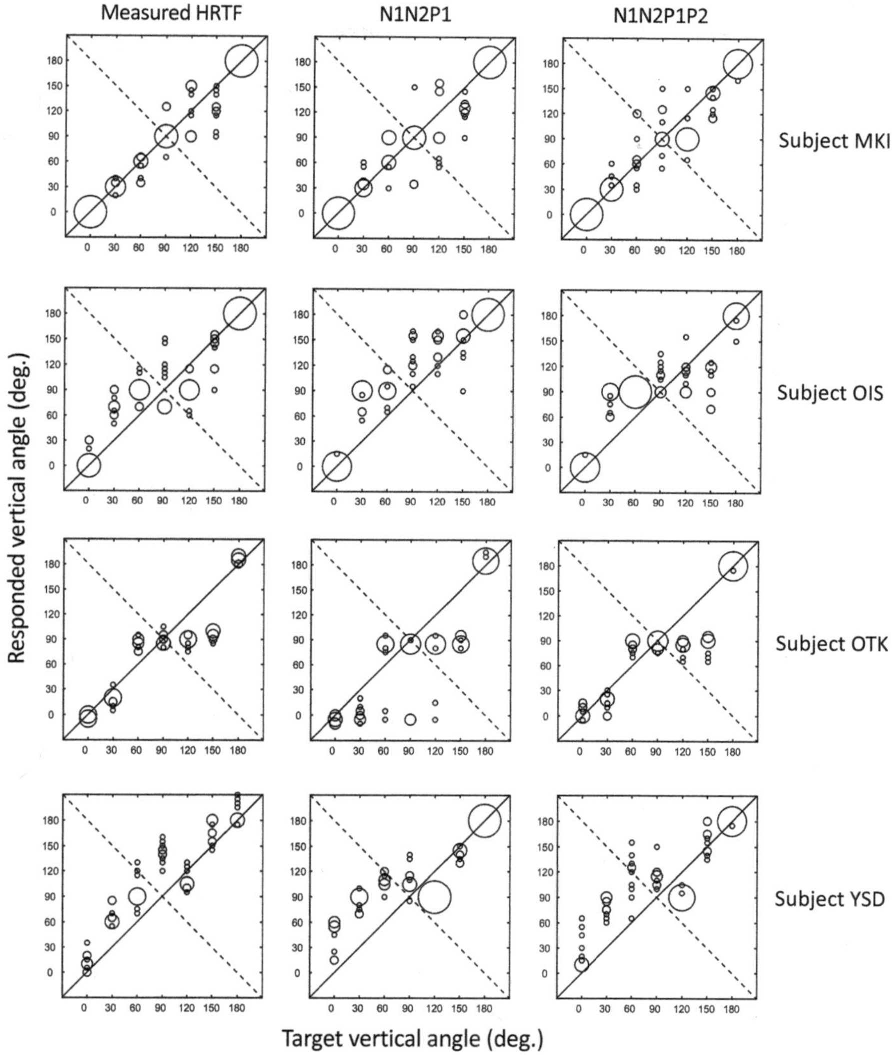

Fig. 3.11 Responses for measured HRTF, parametric HRTF(N1N2P1), and parametric HRTF (N1N2P1P2). (Iida and Ishii 2018)

As a side note, a sound image was not perceived in the upper direction when only P1, P2, or P1P2 were reproduced (Fig. 3.12). These results imply that P1 and P2 were not sufficient in themselves for localization of the upper direction.

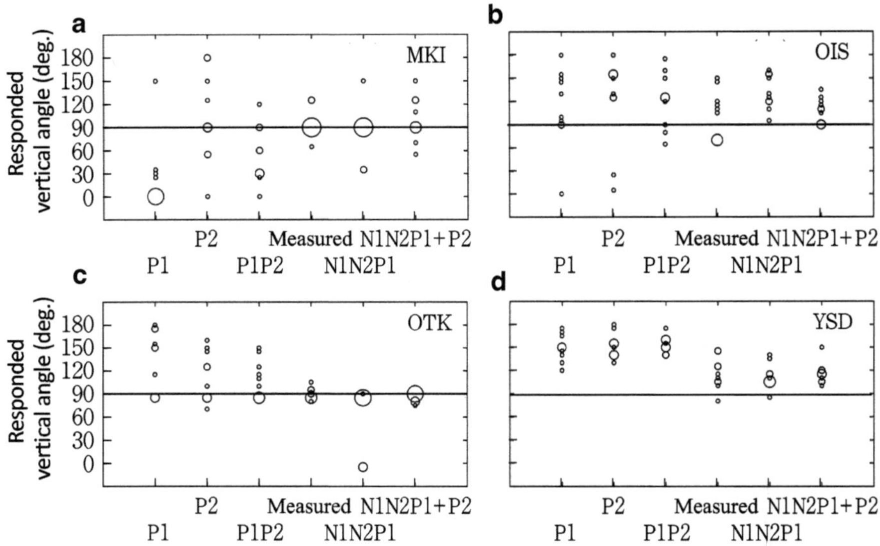

Fig. 3.12 Responses for P1, P2, and P1P2. For comparison, the responses for the measured HRTF, N1N2P1, and N1N2P1P2 in Fig. 3.11 are also shown. (**a**) subject MKI, (**b**) subject OIS, (**c**) subject OTK, and (**d**) subject YSD. (Iida and Ishii 2018)

2. **Vertical angle dependence of notch frequency**

This section examines the relationship between the vertical angle of a sound source and the frequencies of N1 and N2. The frequencies of N1 and N2 strongly depend on the vertical angle of the sound source (Fig. 3.13).

The N1 frequency increases with increasing vertical angle of the sound source from 0° to 120° and then decreases toward 180°. The N2 frequency increases with increasing vertical angle from 0° to 120°, whereas the range of the change in frequency between 120° and 180° is small.

This behavior explains the reason why two notches are necessary for median plane localization. If the notch frequency changed monotonically with the vertical angle of a sound source, then the vertical angle could be determined by extracting only one notch from the ear-input signal. However, since the relationship between the notch frequency and the vertical angle of a sound source is not one-to-one, at least two notches are necessary to determine the vertical angle.

The results of experiments using wide-band white noise with a notch of 8 kHz demonstrated that the subjects were able to detect the notch in the HRTFs and discriminate the difference in frequency of the notch (Moore et al. 1989). These results support the hypothesis that N1 and N2 are the cues for median plane localization.

3. **Vertical angle dependence of peak frequency**

On the other hand, the frequencies of P1 and P2 are almost constant, independent of the vertical angle. Therefore, the physical cue, which depends on the sound source direction is not included in P1 or P2.

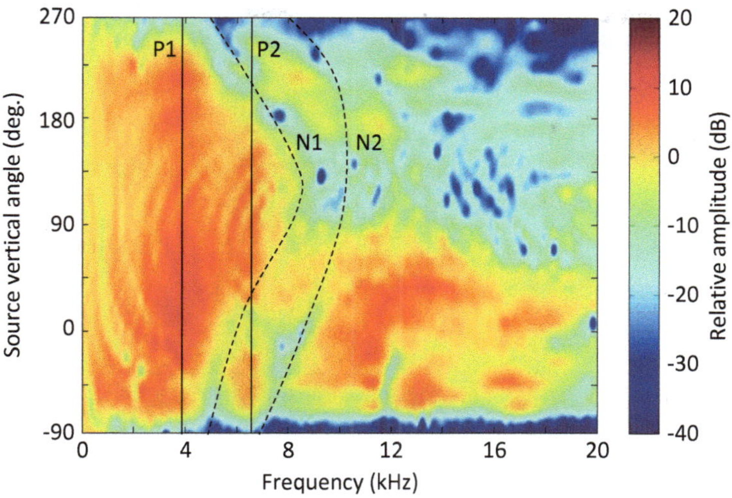

Fig. 3.13 Relationship between vertical angle of sound source and frequencies of N1, N2, P1, and P2

Here, the question is "Why are P1 and P2, the frequencies of which are independent of the vertical angle of a sound source, effective for median plane localization?". The following are possible roles that P1 and P2 play.

One possible interpretation is that the human hearing system uses P1 and P2 as reference information to search for N1 and N2. A listener hears not the HRTFs, but rather the ear-input signals. The ear-input signals are a convolution of the source signal, the spatial (room) impulse response, and the HRIR (see Appendix A.2). Furthermore, the sound pressure level of the ear-input signals varies hour to hour, and the ear-input signals often include background noise.

For extracting N1 and N2 from such ear-input signals, P1 and P2, the frequencies of which are independent of the vertical angle of the sound source, are considered to provide useful reference information to detect N1 and N2 for the human hearing system.

Another possible interpretation is that P1 and P2 emphasize N1 and N2. Figure 3.14 shows the relationship among the frequencies of N1, N2, P1, and P2 and the levels for the vertical angles of 0° (front) and 90° (above). The white and black circles indicate the results for 0° and 90°, respectively.

The two dashed lines indicate the maximum and minimum values of the notch detection threshold for three subjects for the notch having a center frequency of 8 kHz and a band width of 25% of the center frequency (Moore et al. 1989). None of the subjects could detect the notch when its level was higher than −9 dB, whereas they could all detect the notch when its level was less than −20 dB.

For 0°, both N1 and N2 were detectable. For 90°, however, N1 was undetectable, and N2 was detectable for some listeners and undetectable for other listeners. The notch and the peak are located in the order of P1, P2, N1, and N2 from the lower frequency for 90°. Therefore, for the case in which P2 is not reproduced, the contrast

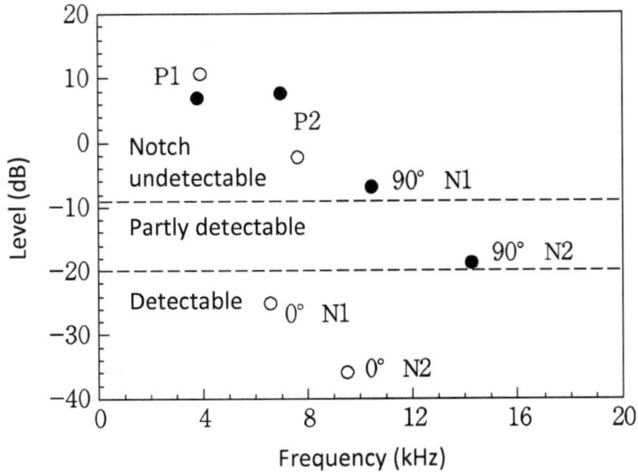

Fig. 3.14 Relationship among frequencies and levels for N1, N2, P1, and P2. Open circles and filled circles indicate the results for 0° and 90°. Broken lines indicate the detection threshold for the notch having a center frequency of 8 kHz and a bandwidth of 25% of the center frequency. (Iida and Ishii 2018)

effects, which emphasize N1, cannot be expected because P1 is at a frequency far from N1. However, for the case in which P2 is reproduced, the relative level of N1 measured from P2 reaches −14.7 dB. At this level, some listeners could detect the notch.

These considerations suggest that P2 could help to improve the accuracy of localization for the upper direction in the median plane by enhancing N1.

3.4 Role of Spectral Information at both Ears in Median Plane Localization

Previous studies clarified that the distortion of the spectral information at one of the ears by occluding the pinna cavities decreases the accuracy of vertical localization (Gardner and Gardner 1973; Morimoto 2001). However, the results of these studies cannot determine whether vertical localization can be accomplished using the spectral information at only a single ear or requires the spectral information at both ears.

The following two hypotheses may explain how the spectral information of the two-ear input signals is processed for vertical angle perception, in other words, the process to extract cues for vertical angle perception from the two ear-input signal spectra.

Hypothesis 1: Spectral cues are extracted from the integrated spectra of the input signals to both ears

The spectra of the two ear-input signals are integrated into one spectrum, and then spectral cues are extracted from the signals and the vertical angle is perceived.

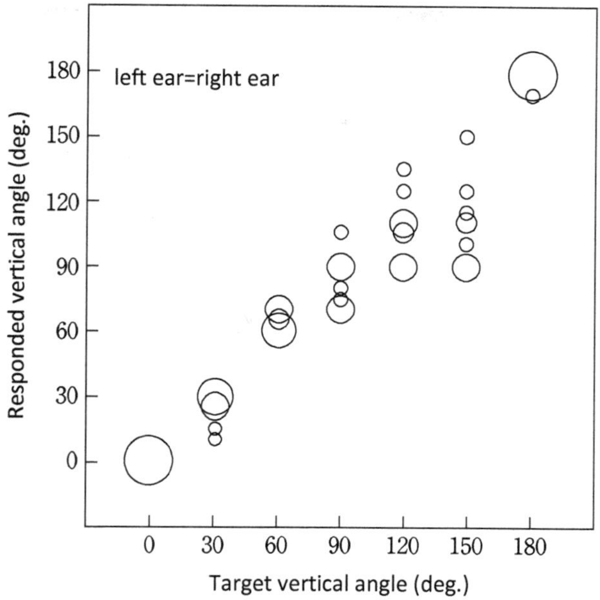

Fig. 3.15 Response when target vertical angles to both ears are identical. (Iida et al. 2018)

Hypothesis 2: Spectral cues are independently extracted from the input signals to each ear

Spectral cues are extracted from the spectrum of input signals to each ear (monaural spectrum) and vertical angle is perceived using these multiple cues.

In order to clarify which hypothesis is valid, median plane localization tests, in which the HRTFs for different target vertical angles were presented to each ear of the listeners, were carried out (Iida et al. 2018).

Figure 3.15 shows the responses for the vertical angle when the HRTFs for identical vertical angles were presented to both ears as usual median plane localization tests. The responses are distributed along a diagonal line, indicating that the subjects perceived the vertical angle of a sound image accurately.

Next, Fig. 3.16 shows the results for the case in which different target vertical angle of HRTFs were provided to the right and left ears. Figure 3.16(a) shows ten responses for the same listener for the case where HRTFs of 0° and 180° were provided to the left and right ears, respectively.

The results showed that the listener either localized a single sound image to the target vertical angle presented to either the left or the right ear, or localized two sound images to both target vertical angles.

Figure 3.16(b) shows the results for the case where HRTFs of 30° and 60° were provided to the left and right ears, respectively. The sound image was perceived at the target vertical angles provided to the left or right ear, and in the intermediate direction only once.

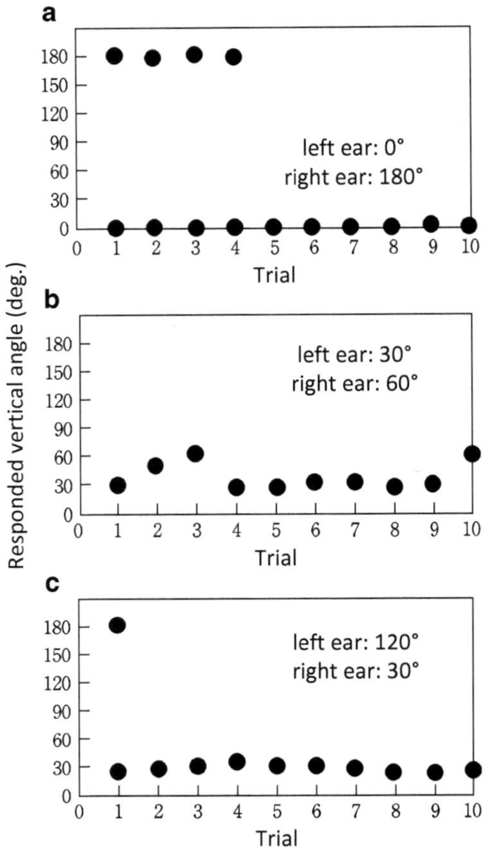

Fig. 3.16 Response when target vertical angles to left and right ears are different

Table 3.1 Mean localization error in the vertical angles for stimuli in which identical target vertical angles were presented to both ears ($\beta_l = \beta_r$) and for stimuli in which different target vertical angles were presented to each ear ($\beta_l \neq \beta_r$). (Iida et al. 2018)

Left ear = Right	Left ear ≠ Right
11.6°	9.3°

Figure 3.16(c) shows the results for the case where HRTFs of 120° and 30° were provided to the left and right ears, respectively. The listener localized most of the sound images to the vertical angle presented to the right ear. However, the listener localized two sound images to both target vertical angles only once.

Table 3.1 shows the mean localization error of the vertical angles for stimuli in which identical target vertical angles were presented to both ears ($\beta_l = \beta_r$) and for stimuli in which different target vertical angles were presented to each ear ($\beta_l \neq \beta_r$). The smaller error for the two target vertical angles was adopted for the stimuli $\beta_l \neq \beta_r$. Both errors were obtained when the subjects localized two sound images. The mean localization error for $\beta_l \neq \beta_r$ was less than that for $\beta_l = \beta_r$. This is probably due to adopting the smaller error in the case of stimuli $\beta_l \neq \beta_r$.

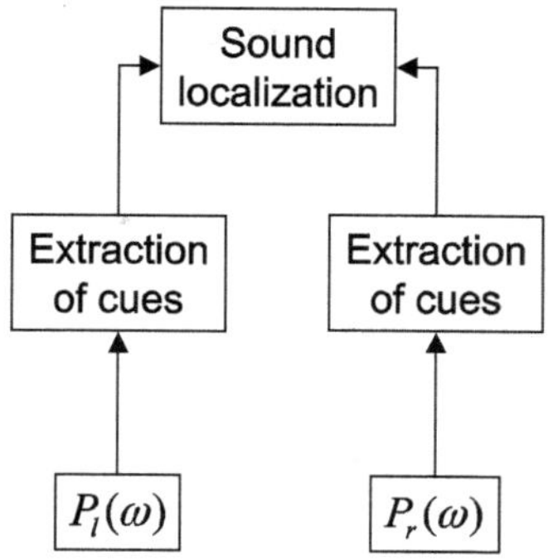

Fig. 3.17 Possible extraction process for spectral cues in human auditory system. The spectral cues are independently extracted from the spectrum of the input signal to each ear, $P_l(\omega)$ and $P_r(\omega)$. (Iida et al. 2018)

A t-test was performed in order to determine whether the difference in the localization error between $\beta_l = \beta_r$ and $\beta_l \neq \beta_r$ was statistically significant. There was no statistically significant difference between the localization errors for $\beta_l \neq \beta_r$ and $\beta_l = \beta_r$.

These results showed that the subjects localized a single sound image to the target vertical angle presented to either the left or right ear or localized two sound images to both target vertical angles, when different target vertical angles were presented to the left and right ears. This implies that a listener perceives the vertical angle of a sound image with the spectral information at only a single ear.

Based on the results, a possible extraction process for spectral cues in the human auditory system is that spectral cues are extracted from the spectrum of the input signal to each ear independently (Fig. 3.17). However, it is not clear which of the ears is dominant in the extraction process. Moreover, what determines this dominance remains unknown.

3.5 Origin of Spectral Cues

We next consider how and where such spectral cues are generated.

3.5.1 Contribution of Pinnae

The HRTFs differ for different sound source directions because of the asymmetry of the pinna, head, and torso in the front-back, up-down, and left-right directions. In particular, the pinnae have the greatest impact on the HRTFs. The pinna is a fold of skin that is supported by cartilage and is 5 to 7 cm in length and 3 to 3.5 cm in width. The pinnae have many cavities and elevated areas, as shown in Fig. 3.18.

Now, which part of the pinnae contributes to the generation of the notches and peaks, and what is the mechanism? A number of studies have examined these problems. Figure 3.19 shows HRTFs calculated by the FDTD method using the

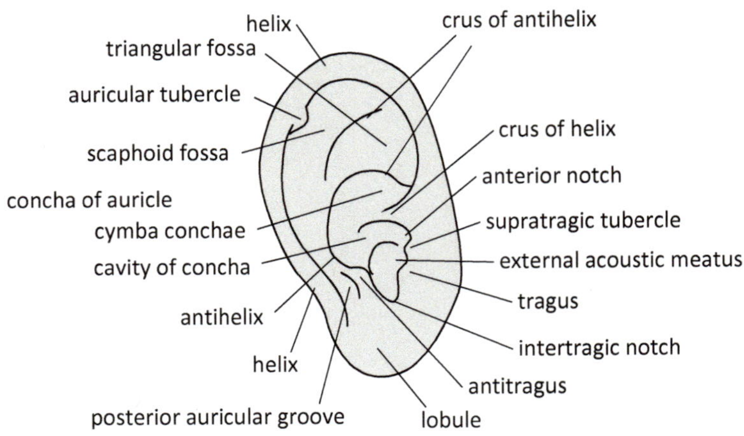

Fig. 3.18 Anthropometry of pinnae

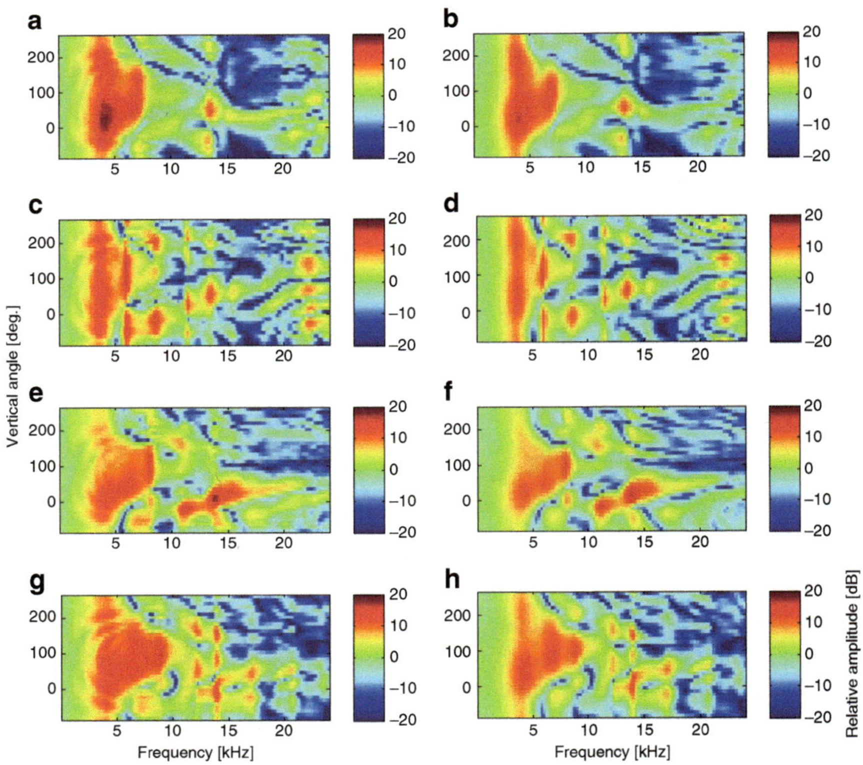

Fig. 3.19 Transfer functions calculated using shape of entire head of subjects M1, M2, F1, and F2 (a, c, e, g) and functions calculated using only shape of pinnae (b, d, f, h). (Takemoto et al. 2012)

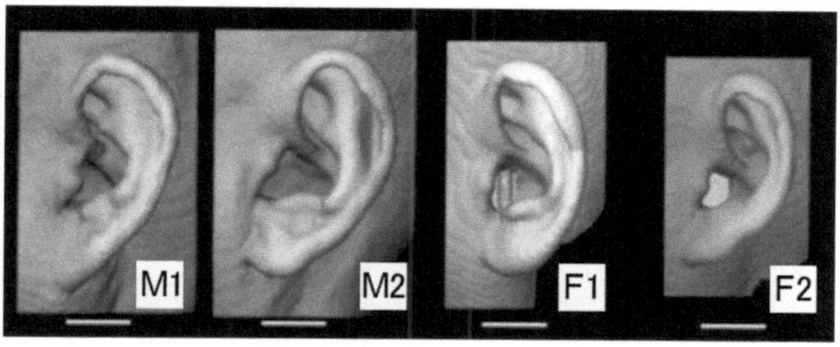

Fig. 3.20 Shape of the pinnae of the four subjects (M1, M2, F1, and F2). The white scale bars indicate 2 cm. (Takemoto et al. 2012)

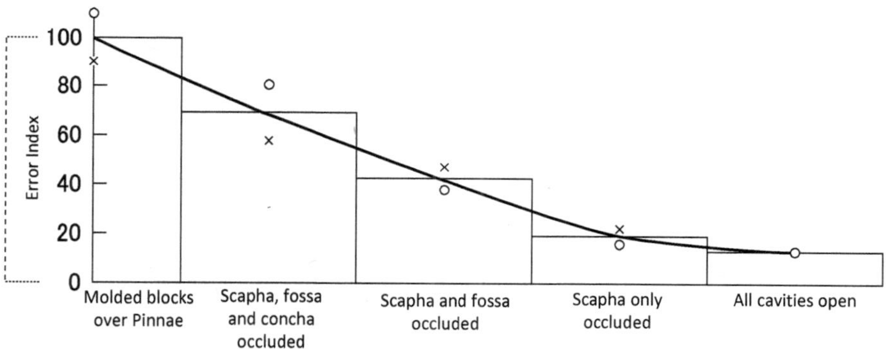

Fig. 3.21 Effects of pinna cavity occlusion on localization in anterior sector of median plane. (Gardner and Gardner 1973)

shape of the entire head of the subjects (a, c, e, g), and HRTFs calculated using only the shape of the pinnae (b, d, f, h) (Takemoto et al. 2012), as shown in Fig. 3.20.

The effects of head shape mainly appear at frequencies lower than 5 kHz. For the case in which the head is considered (a, c, e, g), a region of high sound pressure appears in a concentric pattern with its center at a vertical angle of 90°. This pattern appears to be generated by diffraction of the sound wave around the head. However, no effect of the head on the prominent peaks and notches, namely P1, P2, P3, N1, and N2, is observed. Notches and peaks of the HRTFs in the median plane are determined not by the head, but rather by the pinnae.

We next present experimental results, which verified the effects of the pinnae on the perception of the vertical angle of a sound image. Figure 3.21 shows the error index of the median plane localization for the case in which the three main cavities (triangular fossa, scaphoid fossa, and cavity of concha) of the pinnae were occluded with rubber one by one, while the entrances of the ear canals were open (Gardner and Gardner 1973). The error index increased as the cavities were occluded.

The HRTFs when these three cavities were occluded were also measured (Iida et al. 1998). The amplitude spectra are shown in Fig. 3.22. Figure 3.23 shows the

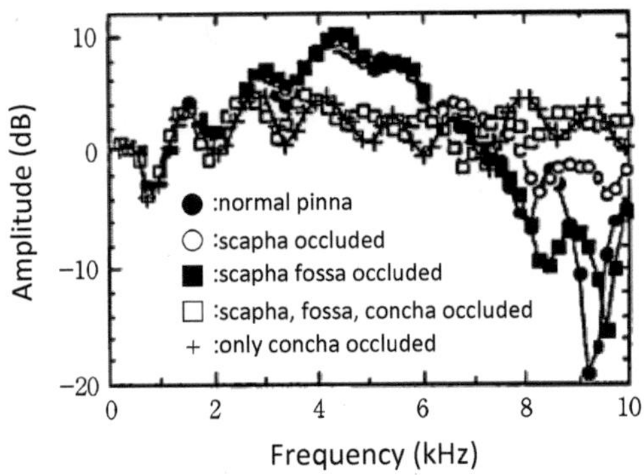

Fig. 3.22 HRTFs of occluded pinnae. (Iida et al. 1998)

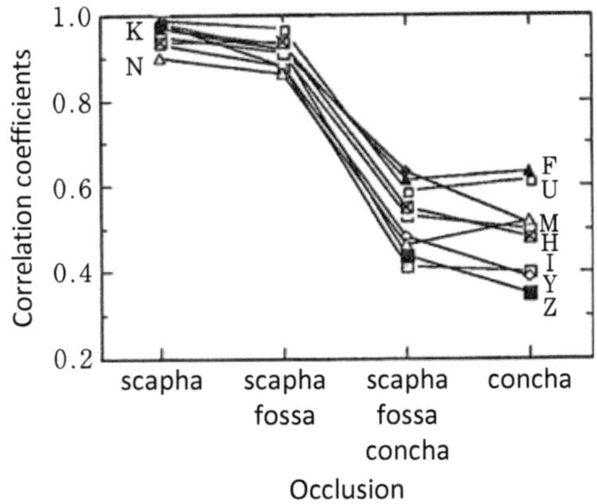

Fig. 3.23 Correlation coefficients between amplitude spectra of HRTFs of occluded pinnae and those of normal pinnae for nine subjects

correlation coefficients between the amplitude spectra of the HRTFs of occluded pinnae and those of the normal pinnae for each of the nine subjects.

Occlusion of the scapha had little effect on the amplitude spectrum, and high correlation coefficients over 0.90 were obtained for all subjects. The notches and peaks for the pinnae, the scapha and fossa of which were occluded, were approximately the same as those for the normal pinnae. The correlation coefficients were over 0.86. For the pinnae, the scapha, fossa, and concha of which were occluded, the notches and the peaks vanished, and the amplitude spectrum was flat. The correlation coefficients were low. In the case that only the concha was occluded, the notches and peaks vanished, and the correlation coefficients were low.

These results suggest that the concha are important for generating the prominent notches and peaks in HRTFs.

Furthermore, sound image localization tests were performed in the upper median plane for seven target vertical angles (0° to 180°, in 30° steps). Table 3.2 shows the number of directions for which a significant difference ($p < 0.01$) was observed in the localization accuracy between the occluded pinnae and the normal pinnae.

Table 3.2 Number of directions for which a significant difference in the localization accuracy was observed compared with the normal pinnae (total of seven directions). (Iida et al. 1998)

	Subject								
Conditions of pinna occlusion	U	M	Z	Y	F	I	K	H	N
Scapha	0	0	1	0	1	0	1	0	0
Scapha+Fossa	0	1	2	1	1	4	3	0	0
Scapha+Fossa+concha	4	4	5	4	4	5	5	0	0
Concha	5	4	3	3	5	4	6	2	1

For the pinnae, the scapha of which was occluded, no significant difference was observed between the normal pinnae and the occluded pinnae in six out of nine subjects for all directions. A significant difference was observed only in one direction for the remainder of the subjects. No tendency for a significant difference for a specific direction was observed.

For the pinnae, the scapha and fossa of which were occluded, a significant difference was observed for six subjects in one to four directions. The subjects perceived a sound image in specific directions (one subject perceived a sound image at around 75° and the other subject at 0°) regardless of the sound source direction).

For the pinnae, the scapha, fossa, and concha of which were occluded, a significant difference was observed for seven subjects in more than four directions. They perceived a sound image in specific directions (one subject perceived a sound image at around 75° and the other subjects perceived a sound image at 0°) regardless of the sound source direction. However, for subjects N and H, no significant difference was observed for any of the sound source directions. The accuracy of localization for the occluded pinnae was approximately the same as that for the normal pinnae. It is not clear which cues subjects N and H used to perceive the vertical angle of a sound image.

For the pinnae, only the concha of which was occluded, a significant difference was observed for seven subjects in more than three directions, as well as the pinnae, the scapha, fossa, and concha of which were occluded. For subjects N and H, a significant difference was observed in one or two directions.

The above results suggest that the cavity of the concha contributes to the generation of the notches and the peaks of HRTFs and significantly affects the perception of the vertical angle of a sound image.

3.5.2 Origin of Peaks

Experiments using a physical pinna model (Shaw 1997) and a simulation using the BEM (Kahana and Nelson 2006) and FDTD methods (Takemoto et al. 2012) revealed that the origin of the peaks is a resonance mode in the pinnae. These studies analyzed the distribution of the antinode of the sound pressure, intensity, and phase difference at the frequencies of the peaks.

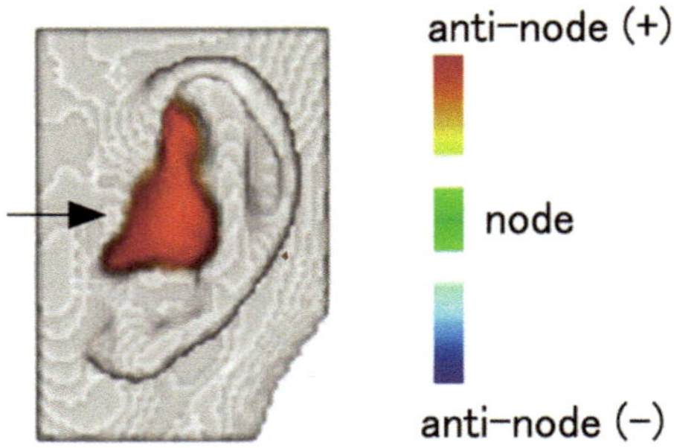

Fig. 3.24 Distribution of sound pressure in pinna at P1 frequency (3.5 kHz) for sound source direction of 0°. (Takemoto et al. 2012)

The results showed that an antinode is generated in the cavity of concha at the peak frequencies. An antinode is generated in the cavity of concha at the P1 frequency. At the P2 frequency, one antinode is generated in the cavity of the concha, and another antinode is generated in the upper cavities of the pinna. At the P3 frequency, one antinode is generated in the cavity of the concha, and two additional antinodes are generated in the upper cavities of the pinna.

Since the antinodes are located in the pinna cavities vertically, these modes are referred to as vertical modes. The origin of each peak is explained in detail below.

1. **First peak (P1)**

The P1 is the first mode generated in the depth direction of the concha. Therefore, the P1 frequency corresponds to the inverse of the wavelength, which is equal to one-fourth the depth of the concha cavity.

Figure 3.24 shows the distribution of sound pressure in the pinna at the P1 frequency (3.5 kHz) for the front direction (0°). An arrow indicates the direction of a sound source. The antinode (+) and antinode (−) indicate high absolute values of the sound pressure, the signs of which are positive and negative, respectively. The node indicates the low absolute value of the sound pressure. The figure shows that an antinode is generated over the entirety of the pinna cavities.

2. **Second peak (P2)**

The origin of P2 is the first mode in the vertical direction in the pinna cavities. This mode is generated along the pinna surface. Two antinodes, the phases of which are reversed, are generated around the entrance of the ear canal and the upper part of the pinna cavities.

Figure 3.25 shows the distribution of sound pressure in the pinna at the P2 frequency (6 kHz) for the upper direction (90°). The antinode of the cavity of concha and that of the cymba conchae and triangular fossa are reversed phase.

Qualitatively, P2 can be described as natural resonance in a rectangular solid room as:

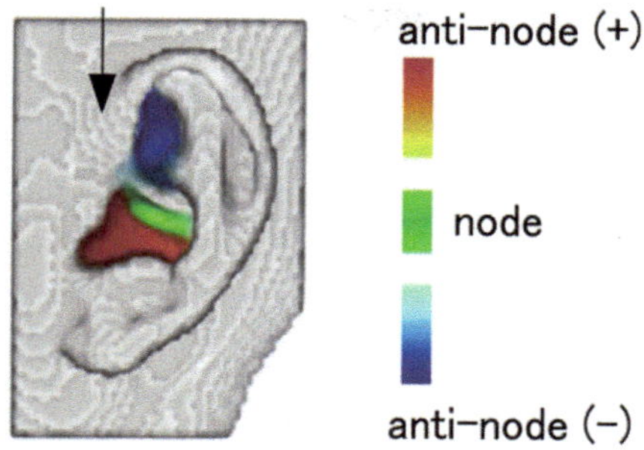

Fig. 3.25 Distribution of sound pressure in pinna at P2 frequency (6 kHz) for sound source direction of 90°. (Takemoto et al. 2012)

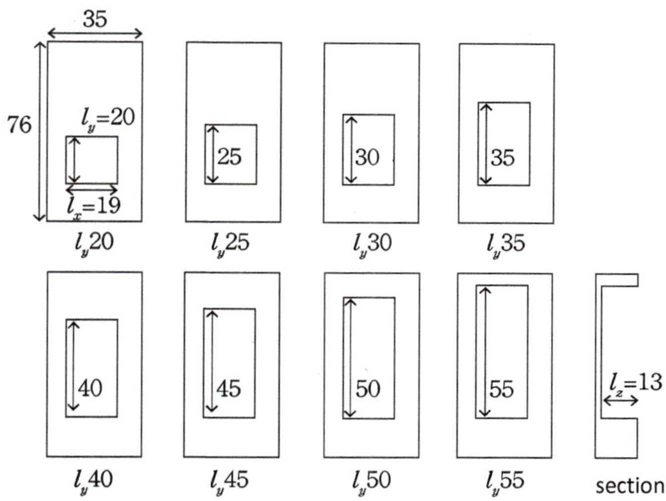

Fig. 3.26 Rectangular solid pinnae model

$$f_n = \frac{c}{2}\sqrt{\left(\frac{n_x}{l_x}\right)^2 + \left(\frac{n_y}{l_y}\right)^2 + \left(\frac{n_z}{l_z}\right)^2}, n_x, n_y, n_z = 0,1,2,\cdots \quad (3.1)$$

where f_n and c denote the natural resonance frequency and the speed of sound, respectively, and l_x, l_y, and l_z indicate the length of each side of the room. Here, P2 corresponds to the first mode for the longest side. (If the longest side is l_x, then, $n_x = 1$, $n_y = n_z = 0$.)

Figure 3.27 shows the measured P2 frequency for the rectangular solid pinnae model shown in Fig. 3.26 and the primary natural resonance frequency calculated using Eq. (3.1). The calculated frequencies show approximately the same behavior

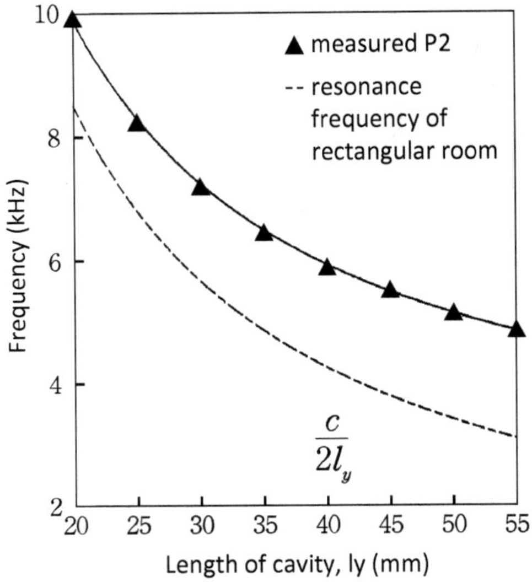

Fig. 3.27 Measured P2 frequency for rectangular solid pinnae model (▲) and calculated primary natural resonance frequency

as that of the measured frequencies. However, the calculated frequencies are lower than the measurement frequencies by approximately 1.5 kHz. The reason for this appears to be that one side of the rectangular solid was open.

3. **Third peak (P3)**

The origin of P3 is the second mode in the vertical direction in the pinna cavities. In this mode, one antinode is generated around the entrance of the ear canal, and two antinodes are generated in the upper cavities of the pinna. Among the two upper antinodes, the antinode close to the entrance of the ear canal has the opposite phase to that of the antinode around the entrance of the ear canal, and the other antinode is in-phase.

Opinion is divided regarding the positions of the two antinodes. It has been reported that the two antinodes are generated in the cymba conchae and the triangular fossa (Shaw 1997), in the cymba conchae and the scaphoid fossa (Kahana and Nelson2006), or in the back side of the cavity of the concha and in the triangular fossa (Takemoto et al. 2012).

The phases and positions of the three antinodes are approximately the same as those of the second natural resonance of a rectangular solid room (Fig. 3.28).

Figure 3.29 shows the distribution of the sound pressure at the P3 frequency (8.25 kHz) for a sound source direction of 120°. The antinode in the concha cavity and the triangular fossa is in-phase, and that at the back side of the concha cavity is reversed phase.

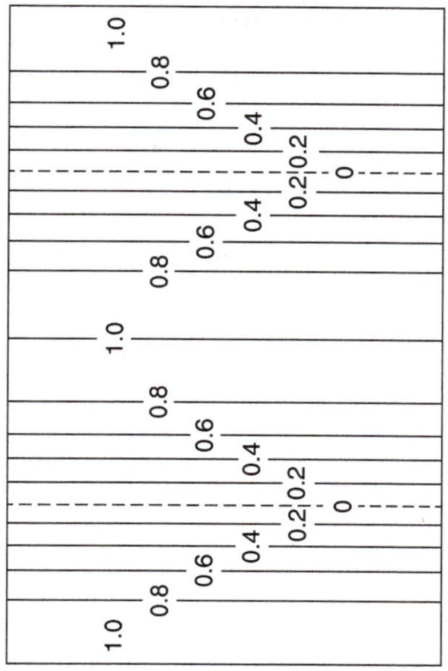

Fig. 3.28 Second natural resonance of rectangular solid room

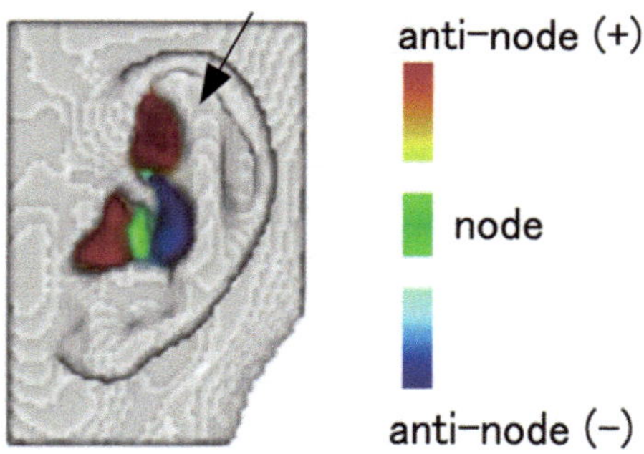

Fig. 3.29 Distribution of sound pressure in pinna at P3 frequency (8.25 kHz) for sound source direction of 120°. (Takemoto et al. 2012)

3.5.3 Origin of Notches

The generation mechanism for the notches is more complicated than that for the peaks, and two hypotheses have been proposed.

One is that the node is generated at the entrance of the ear canal by the interference of the direct wave and the reflected wave from the concha wall (Fig. 3.30) (Raykar et al. 2005). In this hypothesis, the notch frequencies are expressed by the following equation:

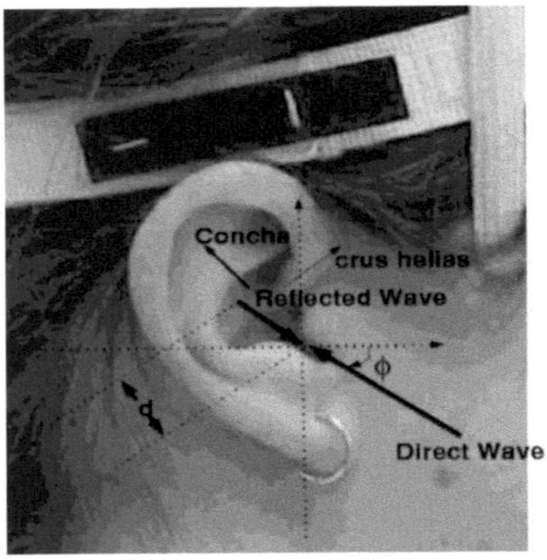

Fig. 3.30 Model of notch generation based on interference of direct wave and reflection from concha wall. (Raykar et al. 2005)

$$f_n(\phi) = \frac{(2n+1)}{2t_d(\phi)}, \quad n = 0,1,2,\cdots \tag{3.2}$$

where f_n is the notch frequency (Hz), t_d is the time difference (s) due to the path length difference between the direct wave and reflected wave, ϕ is the incidence angle of the direct wave, and n is the notch number ($n = 0$ corresponds to the first notch).

However, for a sound source in the upper directions, there exists no reflection point (concha wall). Even for a sound source in the front direction, the notch frequencies calculated by Eq. (3.2) have been reported to not coincide with the measured notch frequencies (Iida et al. 2011).

The other hypothesis is that multiple antinodes with different phases are generated in the pinna, and the node is formed around the entrance of the ear canal (Takemoto et al. 2012). The sound pressure at the entrance of the ear canal becomes a minimum at the N1 frequency.

Figure 3.31 shows the distribution of the sound pressure at the N1 frequencies for six sound source directions. Figure 3.31(a) and (b) show the distribution of the sound pressure for vertical angles of −30° and 0°, respectively. For these angles, N1 appears at 5.25 kHz and 5.5 kHz, respectively. In either case, antinodes are generated in the triangular fossa and the cymba conchae, and a node is generated in the concha cavity. The location of the antinodes and the nodes varies slightly depending on the vertical angle of the sound source.

Figure 3.31(c) and (d) show the distribution of the sound pressure for vertical angles of 30° and 50°, respectively. For these angles, N1 appears at 6.25 kHz and 7.0 kHz, respectively. In this case, two antinodes, the phases of which are reversed, are generated in the triangular fossa, the cymba conchae, and a part of the concha cavity, and a node is generated in the concha cavity.

Figure 3.31(e) and (f) show the distribution of the sound pressure for vertical angles of 150° and 180°, respectively. For these angles, N1 appears at 8.25 kHz and 6.75 kHz, respectively. In this case, two reverse-phase antinodes are generated in the

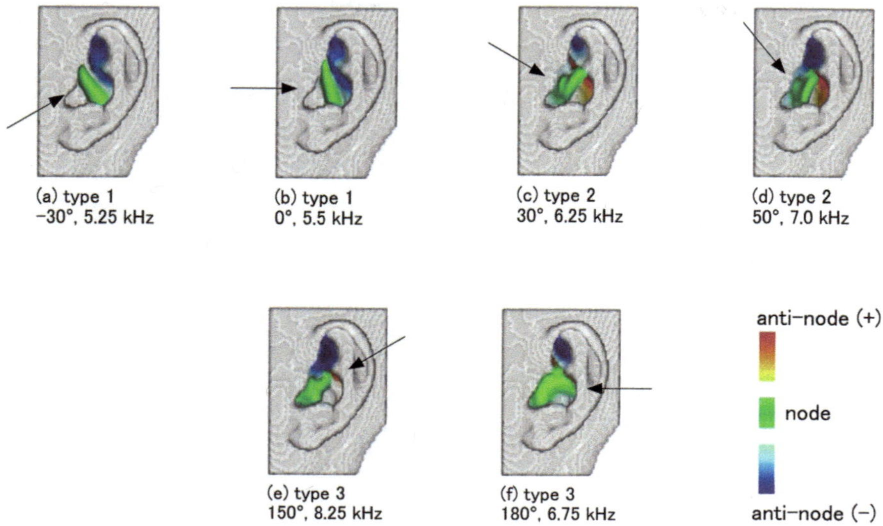

Fig. 3.31 Antinodes and nodes of sound pressure in pinnae. The arrows indicate the direction of the sound source. (Takemoto et al. 2012)

cymba conchae and the triangular fossa, and a node is generated in the region between the two antinodes through the concha cavity.

Numerical calculations on the distribution of the sound pressure around the pinnae, in which a sound source was placed at the entrance of the ear canal, have revealed that nodal lines are generated at specific directions at the N1 frequency. Figure 3.32 shows the sound pressure distributions for the frequencies of (a) 5.75 kHz, (b) 5.95 kHz, and (c) 6.35 kHz. The two straight lines in each figure indicate the vertical angles at which N1 appears. The vertical angles are (a) 19° and 221°, (b) 26° and 216°, and (c) 37° and 186°. These lines coincide with the nodal lines. The vertical angle of the nodal line increases with increasing frequency.

The above observation suggests that two resonances at the same frequency but with opposite phase are generated in the concha cavity and in the area between the cymba conchae and triangular fossa. These resonances then generate the nodal lines. A sound source located on the nodal line gives rise to N1 at the resonance frequency.

3.6 HRTF Learning by Subjects

In order to perceive the direction of a sound image by spectral cues detected from ear input signals, the relationship between the sound source direction and spectral cues must have been acquired by learning.

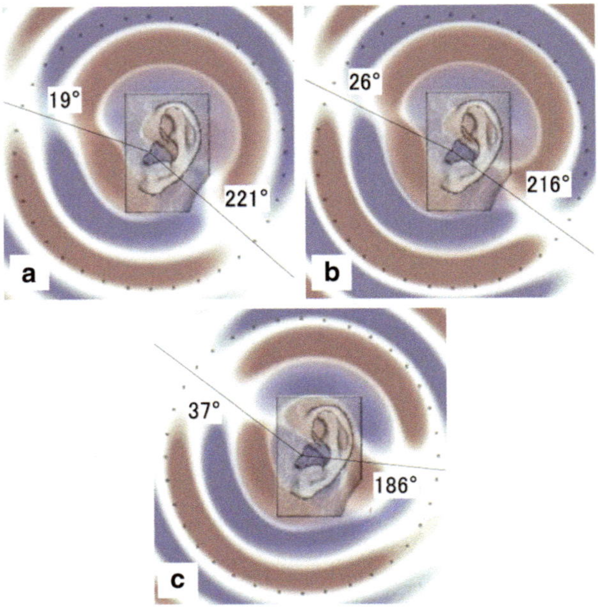

Fig. 3.32 Distribution of instantaneous sound pressure around pinnae. The red and blue regions indicate the local maxima and local minima of the sound pressure, respectively

A study on relearning of the spectral cues by adult subjects has been conducted (Hofman et al. 1998). In the study, the cavities of the pinnae of the adult subjects were occluded with polyester and wax to invalidate the spectral cues the subjects have already acquired. The deterioration of the localization accuracy of vertical directions was confirmed. Then, the subjects were requested to continue daily life under the occluded pinna condition. Three to six weeks were required until the subject was able to achieve accurate sound image localization by relearning spectral cues.

After relearning was accomplished, the fillings were removed, and the pinnae returned to the original state. Interestingly, the subjects localized a sound image accurately both with and without the filling.

This result suggests that the look-up table in the brain, which shows the relationship between the sound source direction and the spectral cues, is not overwritten by relearning, but is newly created.

It is presumed that learning is continuously performed in the developmental stage of the pinnae. The width and length of human pinnae reach their adult size at the age of 3 to 4 years and 9 to 10 years, respectively. Therefore, the learning is considered to be finished in childhood.

3.7 Knowledge of Sound Source

What a listener hears is not HRTFs, but rather ear-input signals. Ear-input signals are expressed, in the frequency domain, by complex multiplication of the spectra of the sound sources, the space transfer functions, and the HRTFs.

As such, the spectra of ear-input signals are not determined by the HRTFs themselves. The question then arises as to whether humans learn not only the spectra of HRTFs, but also the spectra of sound sources. For example, will the sound image localization accuracy differ between sound sources that the subject has heard and sound sources that the subject has never heard? The results of experiments to determine this are shown in Figs. 3.33 and 3.34.

Figure 3.33(a) and (b) show the results of median plane sound image localization tests for the voice of someone with whom the subjects are familiar and for the voice of someone with whom the subjects are not familiar, respectively.

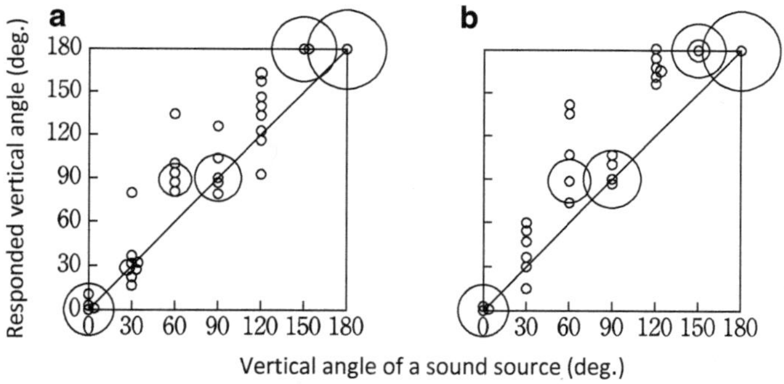

Fig. 3.33 Vertical angle response for (**a**) familiar and (**b**) unfamiliar voices

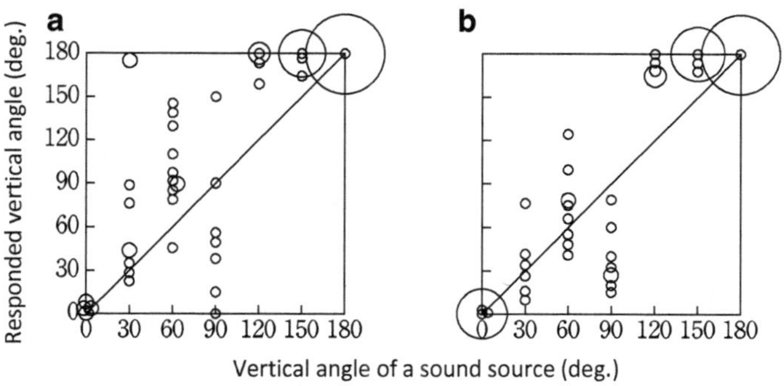

Fig. 3.34 Vertical angle response for (**a**) violin (four seconds) and (**b**) wide-band stationary noise corresponding to average spectrum of violin

Figures 3.34(a) and (b) show the results of median plane sound image localization tests for a violin solo (four seconds) and wide-band stationary noise, the spectrum of which is the same as the average spectrum of the violin solo, respectively.

These results show that the localization accuracy for the unfamiliar sound source is approximately the same as that for the familiar sound source. These results suggest that humans do not learn spectral differences in sound sources.

3.8 Physiological Mechanism of Notch Detection

Does a physiological mechanism to detect notches exist in the human auditory system? Experiments with cats have revealed that the dorsal cochlear nucleus (DCN) distinguishes notches in HRTFs and that Type IV neurons in the DCN extract not the center frequency of the notches, but rather the edges of the high-frequency side of the notches (Reiss and Young 2005).

Furthermore, it has been shown that a function to extract the edges of the high-frequency side is important for vertical angle perception for various kinds of sound sources with different spectra (Baumgartner et al. 2014).

The theory that the edge on the high-frequency side of the notch is important qualitatively supports the statement that an "HRTF of a large pinna tends to be applicable to a listener with small pinnae" in Sect. 2.2.2. Since the notch frequency of the HRTF of a large pinna is lower than that of the small pinna, the edge of the high-frequency side of the notch of the large pinna is included in the notch of the small pinna. On the other hand, the frequency of the edge of the high-frequency side of the notch of the small pinna is higher than that of the large pinna.

3.9 Head Movement

The above sections have discussed vertical perception while keeping the head immobile. Head movement is considered to be another cue for front-back perception. The characteristics of the ear-input signals change according to the head movement of the listener. Suppose the front-back direction of a sound image is uncertain. If the listener turns his/her head to the right and the sound image moves to the left, then the listener judges the sound source to be in the front.

The effects of head movements have already been examined in various sound image localization tests (Perrett and Noble 1997; Kato et al. 2003; Iwaya et al. 2003). Figure 3.35 shows the front-back error rate obtained by sound localization tests, using band-limited noise presented from one of 12 loudspeakers placed in the horizontal plane at an interval of 30° in an anechoic room.

In the figure, the white bars and the black bars indicate the front-back error rates for the restricted head movement and the prompted head movement, respectively. The low-pass noise (LPN) has a cut-off frequency of 1 kHz, and the high-pass noise

Fig. 3.35 Front-back error rates for restricted head movement (white bars) and prompted head movement (gray bars). (Iwaya et al. 2003)

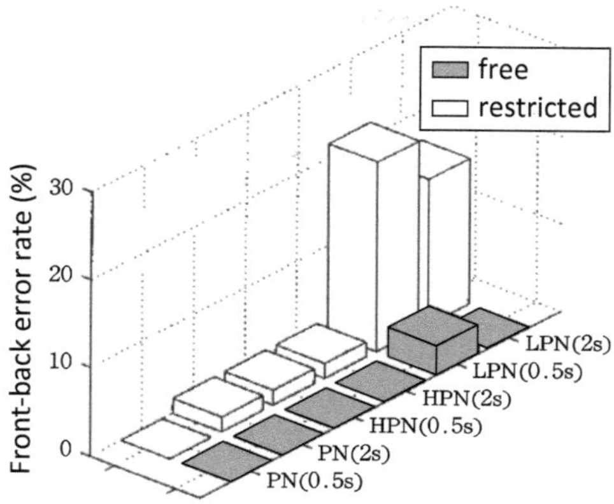

(HPN) has a cut-off frequency of 3 kHz. Finally, pink noise is denoted as PN. The parentheses indicate the duration of the presentation of the stimuli.

For LPN, in which the spectral cues were lost, the front-back error rate was around 20% for the restricted head-movement condition. On the other hand, the rate was less than several percent when the head movement was prompted.

As such, the accuracy of front-back judgment is improved in the experiment in which the subjects were prompted to move their heads. However, humans do not use head movement as cues for front-back judgment in daily life.

Even barn owls, which identify the direction of prey based on the sound emitted by the prey, do not move their heads to obtain cues for localization. Barn owls do not search for a sound source by changing the direction of their head, but rather determine the position of the sound source before moving their head. They then memorize the position of the sound source and turn their heads in the direction of the prey. Similarly, humans are able to perceive the sound direction without moving their head, and moving the head spontaneously to confirm the sound direction is rather unusual (Nojima et al. 2013).

References

Algazi VR, Avendano C, Duda RO (2001) Elevation localization and head–related transfer function analysis at low frequencies. Acoust Soc Am 109:1110–1122

Asano F, Suzuki Y, Sone T (1990) Role of spectral cues in median plane localization. J Acoust Soc Am 88:159–168

Baumgartner R, Majdak P, Laback B (2014) Modeling sound–source localization in sagittal planes for human listeners. J Acoust Soc Am 136:791–802

Butler A, Belendiuk K (1977) Spectral cues utilizes in the localization of sound in the median sagittal plane. J Acoust Soc Am 61:1264–1269

Gardner MB, Gardner RS (1973) Problem of localization in the median plane: effect of pinnae cavity occlusion. J Acoust Soc Am 53:400–408

Hebrank J, Wright D (1974) Spectral cues used in the localization of sound sources on the median plane. J Acoust Soc Am 56:1829–1834
Hofman PM, Van Riswick JGA, Van Opstal AJ (1998) Relearning sound localization with new ears. Nat Neurosci 1:417–421
Iida K, Yairi M, Morimoto M (1998) Role of pinna cavities in median plane localization. In 16th international congress on acoustics, Seattle, 845–846
Iida K, Itoh M, Itagaki A, Morimoto M (2007) Median plane localization using parametric model of the head–related transfer function based on spectral cues. Appl Acoust 68:835–850
Iida K, Gamoh N, Ishii Y (2011) Contribution of the early part of the head-related impulse response to the formation of two spectral notches of vertical localization cues. Forum Acusticum 2011. Aalborg:2241–2245
Iida K, Ishii Y (2018) Effects of adding a spectral peak generated by the second pinna resonance to a parametric model of head-related transfer functions on upper median plane sound localization. Appl Acoust 129:239–247
Iida K, Itoh M, Morimoto M (2018) Upper median plane localization when head-related transfer functions of different target vertical angles are presented to the left and right ears. Acoust. Sci. & Tech. 39:275–286
Iwaya Y, Suzuki Y, Kimura D (2003) Effects of head movement on front-back error in sound localization. Acoust Sci Tech 24:322–324
Kahana Y, Nelson PA (2006) Numerical modelling of the spatial acoustic response of the human pinna. J Sound Vibration 292:148–178
Kato M, Uematsu H, Kashio M, Hirahara T (2003) The effect of head motion on the accuracy of sound localization. Acoust Sci Tech 24:315–317
Kulkarni A, Colburn HS (1998) Role of spectral detail in sound-source localization. Nature 396:747–749
Langendijk EHA, Bronkhorst AW (2002) Contribution of spectral cues to human sound localization. J Acoust Soc Am 112:1583–1596
Lopez–Poveda EA, Meddis R (1996) A physical model of sound diffraction and reflections in the human concha. J Acoust Soc Am 100:3248–3259
Macpherson EA, Sabin AT (2013) Vertical–plane sound localization with distorted spectral cues. Hear Res 306:76–92
Mehrgardt S, Mellert V (1977) Transformation characteristics of the external human ear. J Acoust Soc Am 61:1567–1576
Middlebrooks JC (1992) Narrow–band sound localization related to external ear acoustics. J Acoust Soc Am 92:2607–2624
Middlebrooks JC (1999) Virtual localization improved by scaling non individualized external–ear transfer functions in frequency. J Acoust Soc Am 106:1493–1510
Moore BCJ, Oldfield SR, Doole GJ (1989) Detection and discrimination of spectral peaks and notches at 1 and 8kHz. J Acoust Soc Am 85:820–836
Morimoto M, Saito A (1977) On sound localization in the median plane –Effects of frequency range and intensity of stimuli–. Technical Report of Technical Committee of Psychological and Physiological Acoustics, Acoust. Soc. Jpn.; H-40-1 (in Japanese)
Morimoto M, Ando Y (1980) On the simulation of sound localization. J Acoust Soc Jpn (E) 1:167–174
Morimoto M (2001) The contribution of two ears to the perception of vertical angle in sagittal planes. J Acoust Soc Am 109:1596–1603
Musicant AD, Butler RA (1984) The influence of pinnae–based spectral cues on sound localization. J Acoust Soc Am 75:1195–1200
Nojima R, Morimoto M, Sato H, Sato H (2013) Do spontaneous head movements occur during sound localization? J Acoust Sci Tech 34:292–295
Perrett S, Noble W (1997) The effect of head rotations on vertical plane sound localization. J Acoust Soc Am 104:2325–2332

Raykar VC, Duraiswami R, Yegnanarayana B (2005) Extracting the frequencies of the pinna spectral notches in measured head related impulse responses. J Acoust Soc Am 118:364–374

Reiss LAJ, Young ED (2005) Spectral edge sensitivity in neural circuits of the dorsal cochlear nucleus. J Neuroscience 25:3680–3691

Shaw EAG, Teranishi R (1968) Sound pressure generated in an external–ear replica and real human ears by a nearby point source. J Acoust Soc Am 44:240–249

Shaw EAG (1997) Acoustical features of the human external ear, binaural and spatial hearing in real and virtual environments. Edited by Gilkey RH, Anderson TR. Lawrence Erlbaum Associates, Mahwah, NJ, pp 25–47

Takemoto H, Mokhtari P, Kato H, Nishimura R, Iida K (2012) Mechanism for generating peaks and notches of head–related transfer functions in the median plane. J Acoust Soc Am 132:3832–3841

Chapter 4
Individuality of HRTF

Abstract As described in Chap. 2 and 3, reproduction of the subject's own HRTFs provides accurate sound image localization. On the other hand, using the other's HRTFs causes problems, such as front-back error, rising of a sound image, and inside-of-head localization. This chapter describes in detail past actions for the individualization of HRTFs and the latest findings in this field.

4.1 Individual Differences in HRTFs

Although sound image control and sound field reproduction based on HRTFs have been historically studied for years, they have not been placed into practical use. The main reason is that individual differences in HRTFs have not been overcome. Current acoustic VR, which can present three-dimensional acoustical sensation to only a specific listener, must be evolved into a universal system that can present three-dimensional acoustical sensation to everyone.

First, the degree of individual differences in HRTFs is described from three viewpoints: amplitude spectrum, spectral cues, and interaural difference cues (ITD and ILD).

4.1.1 Individual Differences in Amplitude Spectra

The HRTFs of seven directions in the upper median plane (0°–180°, 30° steps) for ten Japanese adults are shown in Fig. 4.1. Individual differences are small up to around 4 kHz in each direction. On the other hand, above 4 kHz, the frequencies and levels of the peaks and notches and their levels vary widely depending on the listener.

Pinna shapes are supposed to influence the HRTFs above 4 kHz because the wavelength of sounds above 4 kHz is comparable to or smaller than the pinna size. Therefore, individual differences in the amplitude spectrum of the HRTF are assumed to be caused primarily by individual differences in the pinna shape.

K. Iida, *Head-Related Transfer Function and Acoustic Virtual Reality*,
https://doi.org/10.1007/978-981-13-9745-5_4

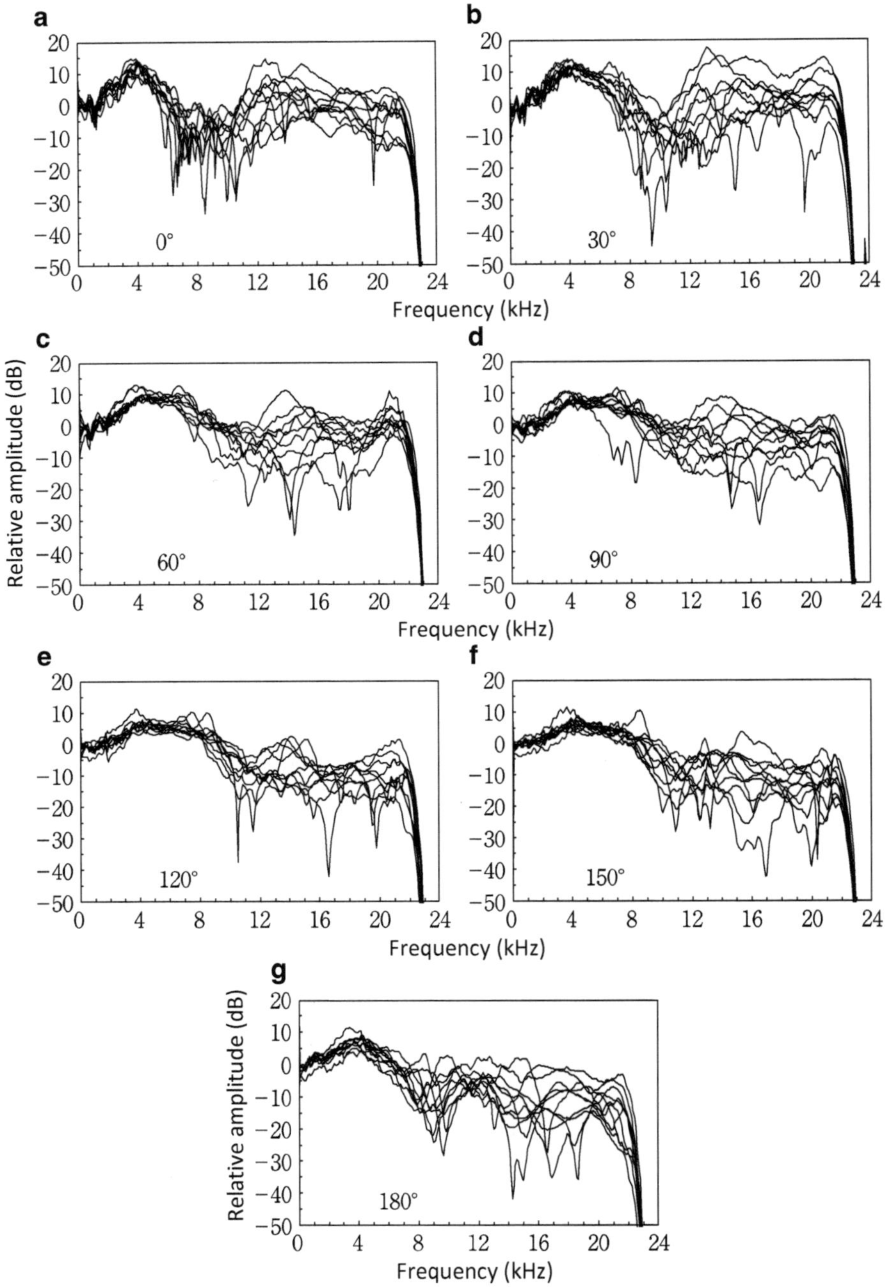

Fig. 4.1 HRTFs of ten Japanese adults in the upper median plane

4.1.2 Individual Differences in Spectral Cues

Individual differences in frequencies of N1 and N2 were analyzed in more detail.

Figure 4.2 shows the distribution of the frequencies of N1 and N2 at seven directions in the upper median plane (0°–180°, 30° steps) for 74 adults (148 ears). The individual differences are summarized in Tables 4.1 and 4.2.

As described in Chap. 3, the N1 frequency increases with an increase in the vertical angle of a sound source from 0° to 120° and then decreases toward 180°. The N2 frequency increases with an increase of the vertical angle from 0° to 120°, while the range of change of the frequency between 120° and 180° is small. The distributions of 0° and 180° almost overlap.

The distributions of N2 frequencies overlap among vertical angles. The N1 and N2 frequencies at a certain vertical angle for a listener may correspond to those at another vertical angle, which differs over 30°, for another listener.

Furthermore, the distribution of the N1 frequency was a normal distribution at all seven directions except 60°. For the N2 frequency, a normal distribution was observed at 0°, 30°, and 90° ($p < 0.05$). However, the distribution was deviated to low frequencies for 60° of N1 and for 60°, 120°, and 150° of N2 and to high frequencies for 180° of N2.

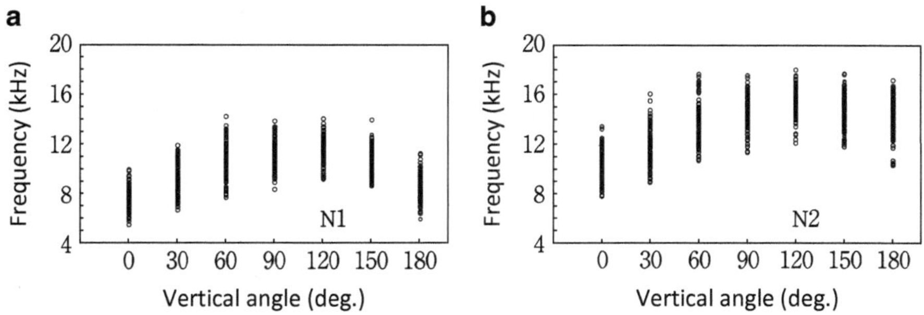

Fig. 4.2 Distributions of (**a**) N1 frequency and (**b**) N2 frequency for 74 adults (148 ears) in the upper median plane

Table 4.1 Individual differences in N1 frequency for 74 adults (148 ears) in the upper median plane

	Vertical angle (deg.)						
N1	0	30	60	90	120	150	180
Ave.	7668	8904	10,369	11,151	11,422	10,702	8600
Max.	9938	11,906	14,250	13,875	14,063	13,969	11,250
Min.	5450	6650	7688	8344	9188	8625	5906
Range (Hz)	4488	5256	6563	5531	4875	5344	5344
Range (oct.)	0.87	0.84	0.89	0.73	0.61	0.70	0.93

Table 4.2 Individual differences in N2 frequency for 74 adults (148 ears) in the upper median plane

	Vertical angle (deg.)						
N2	0	30	60	90	120	150	180
Ave.	10,354	11,806	13,584	14,519	15,147	14,613	14,176
Max.	13,406	16,031	17,625	17,531	18,000	17,719	17,156
Min.	7781	8906	10,688	11,344	12,094	11,800	10,313
Range (Hz)	5625	7125	6938	6188	5906	5919	6844
Range (oct.)	0.78	0.85	0.72	0.63	0.57	0.59	0.73

Table 4.3 Individual differences in P1 frequency for 61 adults (122 ears) in the upper median plane

Ave.	Max.	Min.	Range (Hz)	Range (oct.)
4059	5250	3469	1781	0.60

Individual differences in the N1 frequency at each vertical angle ranges from 0.61 to 0.93 octaves, and that in the N2 frequency ranges from 0.57 to 0.85 octaves. Since the just noticeable difference (JND) of the N1 and N2 frequencies on the vertical angle perception at the front direction is around 0.1–0.2 octaves (Iida and Ishii 2011b), individual differences in the N1 and N2 frequencies are assumed to influence the perception of vertical angle significantly.

On the other hand, the P1 frequency does not depend on the vertical angle of a sound source, as shown in Chap. 3. Table 4.3 shows individual differences in the P1 frequency for 61 adults (122 ears) at the front direction. The individual difference in the P1 frequency is 0.60 octaves. This is smaller than the individual differences in the N1 and N2 frequencies (0.87 and 0.78 octaves, respectively). The JNDs of the P1 frequency on the vertical angle perception at the front direction is 0.35 octaves on the high-frequency side and 0.47 octaves on the low-frequency side (Iida et al. 2014). Therefore, using the median value of the P1 frequency, individual differences in the P1 frequency are assumed to hardly influence perception of vertical angle.

4.1.3 Individual Differences in Interaural Time Difference

Figure 4.3 shows the ITD of 12 directions in the horizontal plane for 33 Japanese adults (27 males, 6 females) (Ishii and Iida 2017). The positive values indicate that the sound wave to the right ear arrives faster than that to the left ear.

The ITD reaches its maximum at the lateral directions (90° and 270°). For some listeners, the ITDs were not 0, even though the direction of the sound source was 0° or 180°. This is due to the asymmetry of the listener's head.

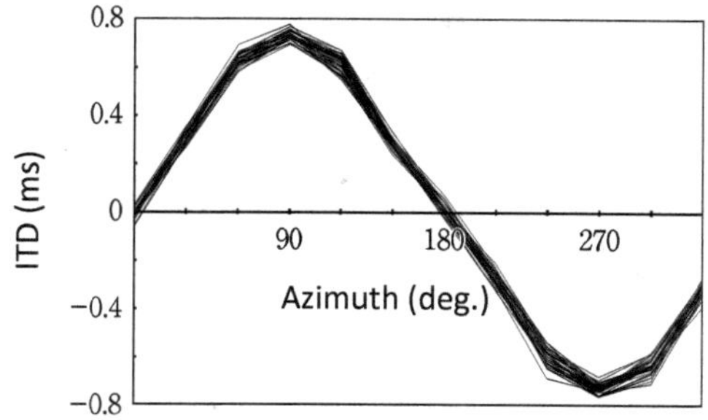

Fig. 4.3 ITD for 33 adults (27 males and 6 females) at twelve azimuth angles (30° steps) in the horizontal plane

Table 4.4 Individual differences in ITD for 33 adults (27 males and 6 females) at twelve azimuth angles (30° steps) in the horizontal plane (μs)

	0°	30°	60°	90°	120°	150°
Max.	26.0	354.2	692.7	778.6	669.3	330.7
Min.	−65.1	265.6	578.1	692.7	546.9	234.4
Range	91.1	88.6	114.6	85.9	122.4	96.3
	180°	210°	240°	270°	300°	330°
Max.	70.3	−210.9	−539.1	−679.7	−580.7	−273.4
Min.	−28.6	−325.5	−687.5	−763.0	−710.9	−393.2
Range	98.9	114.6	148.4	83.3	130.2	119.8

Table 4.4 shows the maximum, minimum, and individual difference (maximum − minimum) of the ITDs. The individual difference in the ITDs at each direction ranges from 83.3 to 148.4 μs. The individual difference was small at the just lateral directions (90° and 270°) and large at 60°, 120°, 240°, and 300° (just lateral ±30° directions). The range of the ITD of 90° overlaps with that of 60°. Furthermore, the range of the ITD of 270° overlaps those for 240° and 300°.

It has been reported that the JNDs of the ITDs for wide-band signals are 19 μs and 72 μs at the front direction and at the lateral direction, respectively (Mills 1958; Hershkowitz and Durlach 1969; Domnitz and Colburn 1977). Therefore, using another subject's ITD, a listener may detect the difference in the azimuth of a sound image.

4.1.4 Individual Differences in Interaural Level Difference

Figure 4.4 shows the ILDs of 12 directions in the horizontal plane for the same 33 Japanese adults (27 males, 6 females) (Ishii and Iida 2017) as those for the ITDs described above. Since the ILD varies depending on frequency, the figure shows the ILDs of five 1/3-octave bands, the center frequency of which was 500 Hz to 8 kHz.

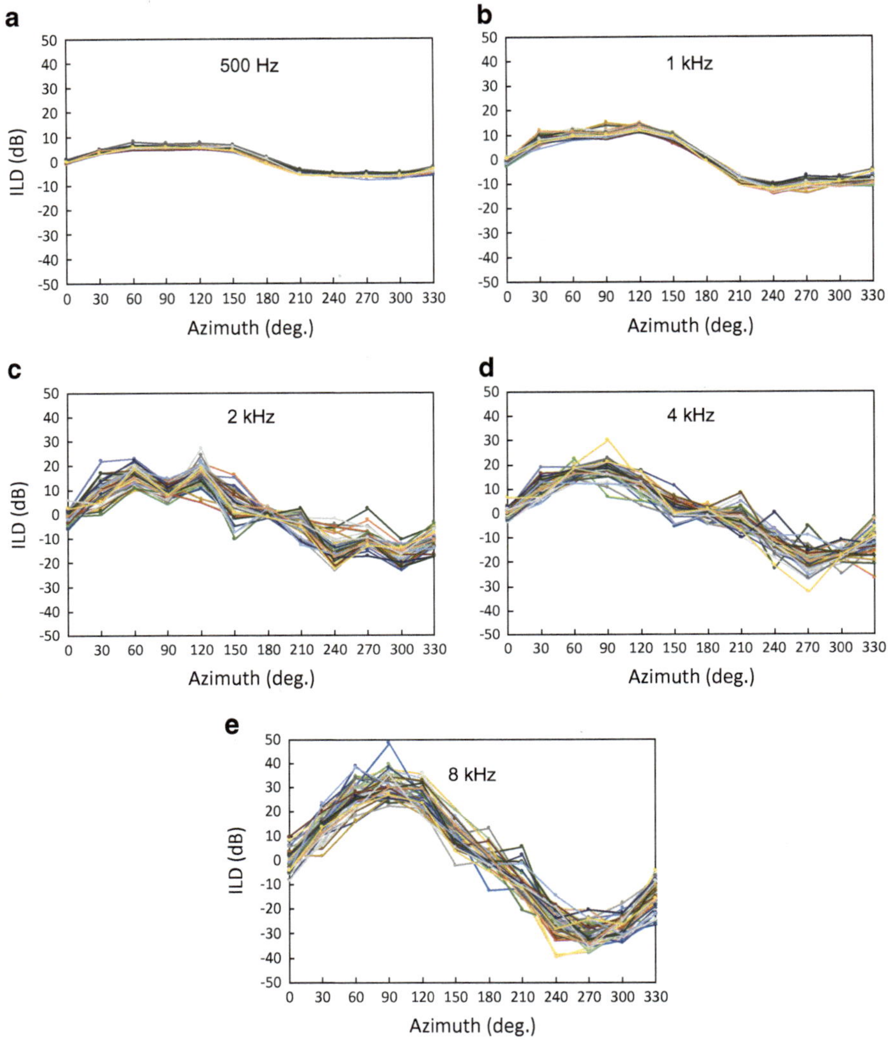

Fig. 4.4 ILD for 33 adults (27 males and 6 females) at twelve azimuth angles (30_ steps) in the horizontal plane for five 1/3-oct. bands. (**a**) 500 Hz, (**b**) 1 kHz, (**c**) 2 kHz, (**d**) 4 kHz, and (**e**) 8 kHz

The positive values indicate that the sound wave to the right ear is stronger than that to the left ear.

The difference in ILD among azimuth angles is within ±10 dB at 500 Hz. On the other hand, for some listeners, the difference reaches around ±30 dB at 8 kHz. In general, the ILD reaches its maximum on lateral sides. However, for the high-frequency range, for some listeners, the ILD at 120° was larger than that at 90°, and the ILD at 240° was larger than that at 270°. This is assumed to be due to the effects of the pinnae.

Tables 4.5, 4.6, 4.7, 4.8 and 4.9 show the maximum, minimum, and individual difference (maximum – minimum) of the ILDs. The individual difference in the ILDs increases as the frequency increases. They range from 1.9 to 3.4 dB at 500 Hz, from 2.8 to 7.2 dB at 1 kHz, from 4.6 to 26.2 dB at 2 kHz, from 8.3 to 27.1 dB at 4 kHz, and from 16.0 to 26.1 dB at 8 kHz.

Table 4.5 Individual differences in ILD for 33 adults (27 males and 6 females) at twelve azimuth angles (30° steps) in the horizontal plane for 1/3-oct. band of 500 Hz (dB)

	0°	30°	60°	90°	120°	150°
Max.	0.9	5.1	8.1	7.2	7.5	6.7
Min.	−1.6	3.0	4.7	4.6	4.9	3.6
Range	2.5	2.1	3.4	2.6	2.7	3.0
Ave.	−0.3	4.2	6.0	5.9	5.9	4.9
	180°	210°	240°	270°	300°	330°
Max.	1.7	−3.4	−4.8	−4.5	−4.6	−2.6
Min.	−1.3	−5.9	−6.6	−7.8	−7.8	−5.7
Range	3.0	2.4	1.9	3.3	3.1	3.2
Ave.	0.3	−4.7	−5.9	−5.9	−6.0	−4.4

Table 4.6 Individual differences in ILD for 33 adults (27 males and 6 females) at twelve azimuth angles (30° steps) in the horizontal plane for 1/3-oct. band of 1 kHz (dB)

	0°	30°	60°	90°	120°	150°
Max.	1.2	11.8	12.8	14.9	14.8	10.7
Min.	−2.8	4.6	8.0	8.3	11.0	6.3
Range	4.0	7.2	4.9	6.6	3.9	4.5
Ave.	−0.5	8.5	10.4	10.1	12.4	8.5
	180°	210°	240°	270°	300°	330°
Max.	1.7	−6.4	−10.0	−7.0	−7.1	−4.2
Min.	−1.0	−10.8	−14.3	−14.0	−11.5	−11.2
Range	2.8	4.5	4.4	7.1	4.5	7.0
Ave.	0.4	−8.2	−12.4	−9.8	−10.0	−8.2

Table 4.7 Individual differences in ILD for 33 adults (27 males and 6 females) at twelve azimuth angles (30° steps) in the horizontal plane for 1/3-oct. band of 2 kHz (dB)

	0°	30°	60°	90°	120°	150°
Max.	5.7	22.0	23.1	14.6	27.2	16.0
Min.	−6.1	0.2	9.8	4.1	5.0	−10.2
Range	11.8	21.8	13.2	10.6	22.2	26.2
Ave.	−0.5	9.1	16.3	9.1	16.1	2.6
	180°	210°	240°	270°	300°	330°
Max.	3.0	2.3	−1.8	2.2	−10.5	−3.6
Min.	−1.7	−12.7	−23.3	−17.3	−23.2	−17.2
Range	4.6	15.0	21.4	19.5	12.7	13.6
Ave.	0.3	−4.1	−14.6	−9.7	−17.3	−10.0

Table 4.8 Individual differences in ILD for 33 adults (27 males and 6 females) at twelve azimuth angles (30° steps) in the horizontal plane for 1/3-oct. band of 4 kHz (dB)

	0°	30°	60°	90°	120°	150°
Max.	6.4	19.2	22.4	30.1	17.5	11.3
Min.	−3.5	4.4	12.5	6.7	3.2	−5.4
Range	9.9	14.8	10.0	23.4	14.3	16.6
Ave.	0.0	10.4	16.6	18.4	11.4	2.4
	180°	210°	240°	270°	300°	330°
Max.	3.9	8.3	0.4	−5.4	−10.9	−1.6
Min.	−4.4	−10.1	−22.7	−32.4	−24.9	−26.6
Range	8.3	18.4	23.1	27.1	13.9	25.0
Ave.	0.0	−2.5	−12.1	−19.7	−16.8	−12.0

Table 4.9 Individual differences in ILD for 33 adults (27 males and 6 females) at twelve azimuth angles (30° steps) in the horizontal plane for 1/3-oct. band of 8 kHz (dB)

	0°	30°	60°	90°	120°	150°
Max.	9.7	23.5	38.8	48.5	35.9	21.1
Min.	−8.6	1.7	15.9	22.5	18.3	−2.3
Range	18.2	21.7	22.9	26.0	17.6	23.4
Ave.	−0.3	14.0	27.1	31.0	25.7	10.5
	180°	210°	240°	270°	300°	330°
Max.	13.1	5.5	−14.7	−20.5	−17.8	−4.5
Min.	−12.5	−20.6	−40.0	−38.0	−33.8	−26.3
Range	25.7	26.1	25.3	17.5	16.0	21.8
Ave.	0.7	−10.0	−25.7	−30.8	−26.6	−15.0

It has been reported that the JND of the ILD on the azimuth direction of a sound image is approximately 1 dB (Hartmann and Constan 2002; Bernstein 2004). Therefore, individual differences in the ILD are assumed to be significant in azimuth perception of a sound image. Accurate reproduction of the ILD is required for control and reproduction of the azimuth angle of a sound image.

4.2 Individual Differences in the Pinna and Head Shape

Individual differences in HRTFs described in the previous section are caused by differences in the pinna and head shape of the listener. Here, individual differences in the pinna and head shape are discussed.

4.2.1 Individual Differences in Pinna Shape

Figure 4.6 and Table 4.10 show the 10 measured anthropometric parameters of the pinnae (see Fig. 4.5) for 111 Japanese adults (222 ears). These parameters were measured from ear molds collected from the subjects beforehand, using a digital vernier caliper.

Here, x_d is the distance to the deepest point (depth max) of the cavum conchae, and x_a (tilt of pinna) was obtained only from 21 subjects (42 ears), using photographs of the profiles of the subjects.

The range of values for each anthropometric parameter spanned 10 to 30 mm, and the angle of tilt, x13, ranged widely from 10° to 40°. Furthermore, the Shapiro-Wilk normality test has shown that each pinna parameter is normally distributed ($p < 0.05$).

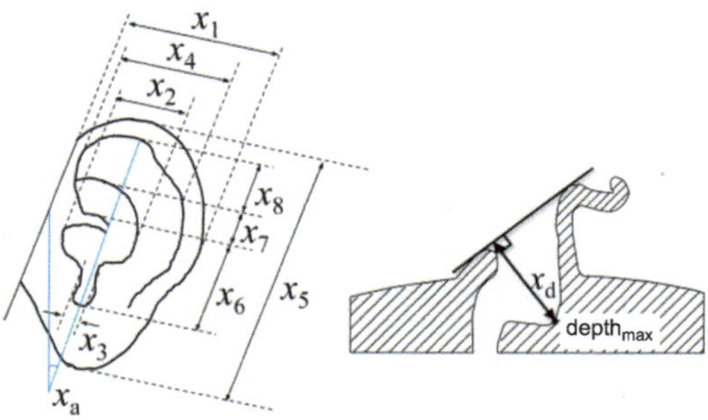

x_1	width of pinna
x_2	width of concha
x_3	width of insisura intertragica
x_4	width of helix
x_5	length of pinna
x_6	length of concha
x_7	length of cymba conchae
x_8	length of scapha
x_d	depth of concha
x_a	tilt of pinna

Fig. 4.5 Ten anthropometric parameters of the pinna

Table 4.10 Individual differences in ten anthropometric parameters of the pinna for 111 Japanese adults (222 ears)

	Ave. (mm)	Max. (mm)	Min. (mm)	σ (mm)	RSD
x_1	34.4	43.4	28.4	2.9	0.08
x_2	19.3	26.1	13.4	2.2	0.11
x_3	8.5	12.0	3.8	1.6	0.19
x_4	25.5	36.8	16.8	3.2	0.13
x_5	65.7	83.2	53.5	4.7	0.07
x_6	21.1	26.7	16.6	1.8	0.08
x_7	5.7	10.3	1.7	1.7	0.30
x_8	17.7	25.0	10.4	2.8	0.16
x_d	13.5	17.7	9.3	1.6	0.12
x_a	25.4°	40.0°	10.0°	8.1°	0.32

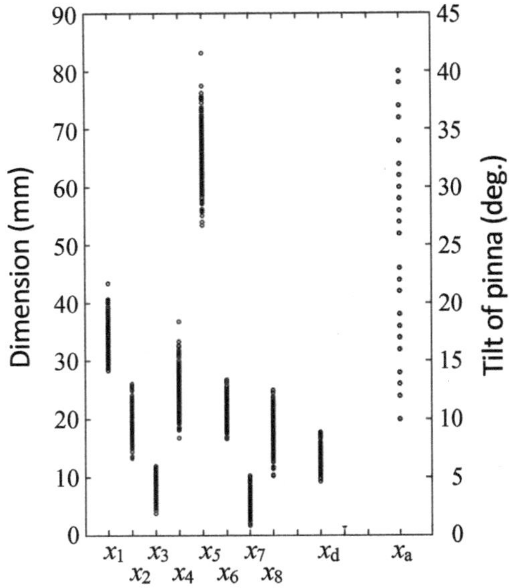

Fig. 4.6 Distribution of ten measured anthropometric parameters of the pinna for 111 Japanese adults (222 ears)

Next, individual differences are discussed based on relative standard deviation (RSD), which is obtained by dividing the standard deviation by the mean value:

$$RSD = \frac{\sigma}{\bar{x}} \tag{4.1}$$

where σ and $\bar{x}$ indicate the standard deviation and the mean value, respectively. This value indicates the relative dispersion of the distribution of the measured parameters.

The RSD of each parameter ranges from 0.07 to 0.32. The RSDs are large for the tilt of the pinna (x_a) and the length of the cymba conchae (x_7). On the other hand, RSDs are small for the maximum ear length (x_5), the maximum ear width (x_1), and the length of the cavum conchae (x_6).

Considering the origin of the spectral notches described in Sect. 3.5.3, the individual difference in x_6 and x_7 may be related to the individual difference of the N1 frequency. However, as mentioned above, the RSD of x_6 is small, whereas that of x_7 is large. Further studies are required to clarify these relationships.

On the other hand, focusing on cavum conchae, which contribute to the generation of P1, the RSDs of the maximum width (x_2), length (x_6), and depth (x_d) of the cavum conchae are small. These correspond to small individual differences in P1 frequency.

4.2.2 Individual Differences in Head Shape

Individual differences in the interaural difference are due primarily to individual differences in the head shape.

A total of 15 anthropometric parameters of the head (see Fig. 4.7) for the same 33 Japanese adults (27 males, 6 females) (Ishii and Iida 2017), as described in Sects. 4.1.3 and 4.1.4, were measured. Here, x_1 and x_5 were measured with a digital vernier caliper, and x_{21}, x_{22}, x_{24}, and x_{29} were measured with by an esthesiometer. In addition, x_{23}, x_{25}, x_{26}, and x_{28} were measured with a tape measure. The reading resolutions are 0.01 mm for the digital caliper and 1 mm for the esthesiometer and tape measure.

Table 4.11 shows the results. The RSDs were one digit smaller than those of pinnae (Table 4.10). In other words, individual differences in the head were remarkably smaller than those of pinnae. The RSDs for pericephalic length (x_{25}, x_{26}) and shoulder length (x_{24}) were relatively large. These parameters appear to be closely related to the ITD and ILD.

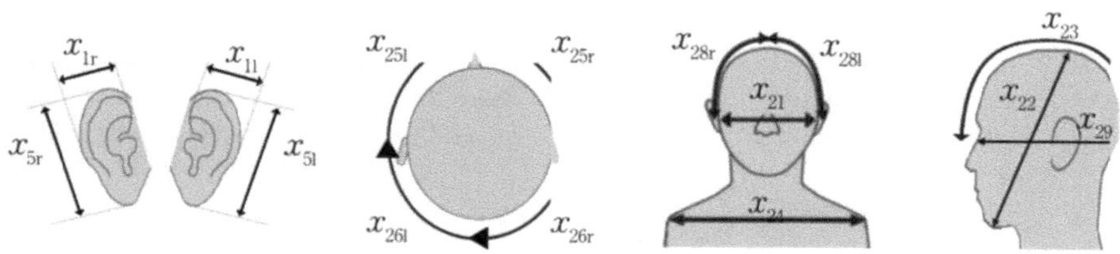

Fig. 4.7 Fifteen anthropometric parameters of the head

Table 4.11 Individual differences in 15 anthropometric parameters of the head for 33 Japanese adults (17 males and 6 females)

	Ave. (mm)	Max. (mm)	Min. (mm)	Range (mm)	σ (mm))	RSD
x_{1l}	31.79	35.32	27.55	7.77	2.1	0.066
x_{1r}	33.47	36.8	27.09	9.71	2.3	0.069
x_{5l}	63.32	72.44	51.4	21.04	4.4	0.070
x_{5r}	62.67	71.38	50.7	20.68	4.5	0.071
x_{21}	143	152	134	18	5.6	0.039
x_{22}	246	266	227	39	9.9	0.040
x_{23}	420	461	386	75	16.6	0.039
x_{24}	393	439	337	102	23.9	0.061
x_{25l}	152	167	136	31	7.4	0.049
x_{25r}	159	185	148	37	8.1	0.051
x_{26l}	144	163	130	33	8.1	0.056
x_{26r}	145	176	120	56	12.0	0.083
x_{28l}	195	217	176	41	8.2	0.042
x_{28r}	197	211	185	26	7.2	0.036
x_{29}	185	201	170	31	7.9	0.043

4.3 Standardization of HRTFs

One possible method by which to solve the individual differences in HRTFs is to find or create “standard HRTFs”. If standard HRTFs, by which everyone can perceive the three-dimensional acoustical sensation, are found or created, the spread of three-dimensional sound systems will progress drastically.

4.3.1 Sound Image Localization with the HRTFs of a Dummy Head

Various kinds of dummy heads have been developed toward such a standardization (e.g. Burkhard and Sachs 1975). However, the dummy heads, which were developed based on the representative shape of the head and pinna obtained from the shape of the head and pinna of numerous listeners, have HRTFs that do not provide accurate sound localization for most listeners.

Figure 4.8 shows the results of median plane sound image localization using the HRTFs of a dummy head (B&K, Type 4128C). A wide-band white noise was presented through the headphones (AKG, K1000), which were regarded as FEC headphones (see Chap. 12), and the transfer functions between the headphones and the entrances of the ear canal of the listeners were compensated in the range of ±1.5 dB.

For subject TCY, most of the responses to the subject’s own HRTFs were distributed along a diagonal line, whereas the variances of the responses were large at 120° and 150°. For the HRTFs of the dummy head, however, the subject never responded at upper directions. The subjects localized a sound image to either the front or rear directions for the target vertical angles of 60° to 120°.

For subject YMM, the responses to the subject’s own HRTFs were distributed as an inverted s-shaped curve centered over a diagonal line. The responses tended to shift slightly upward for the target vertical angles of 0° to 90°. However, for the HRTFs of the dummy head, the subject never responded at the front directions.

For subject OIS, the responses to the subject’s own HRTFs were distributed as an inverted s-shaped curve centered over a diagonal line and tended to shift slightly upward for the target vertical angles of 60° and 120°. For the HRTFs of the dummy head, the subjects distinguished the front, upper, and rear directions of a sound image, although the responses were distributed widely.

Figure 4.9 shows the mean vertical localization error for the subject’s own HRTFs, the HRTFs of the dummy head, and the actual sound sources. For the target vertical angles of 0° and 180°, the mean vertical localization error for the subject’s own HRTFs and the actual sound sources was less than 10°. However, the errors for the HRTFs of the dummy head for target vertical angles of 0° and 180° were 40° and 22°, respectively. For 30° and 150°, there exist no remarkable differences among the subject’s own HRTFs, the HRTFs of the dummy head, and the actual sound sources.

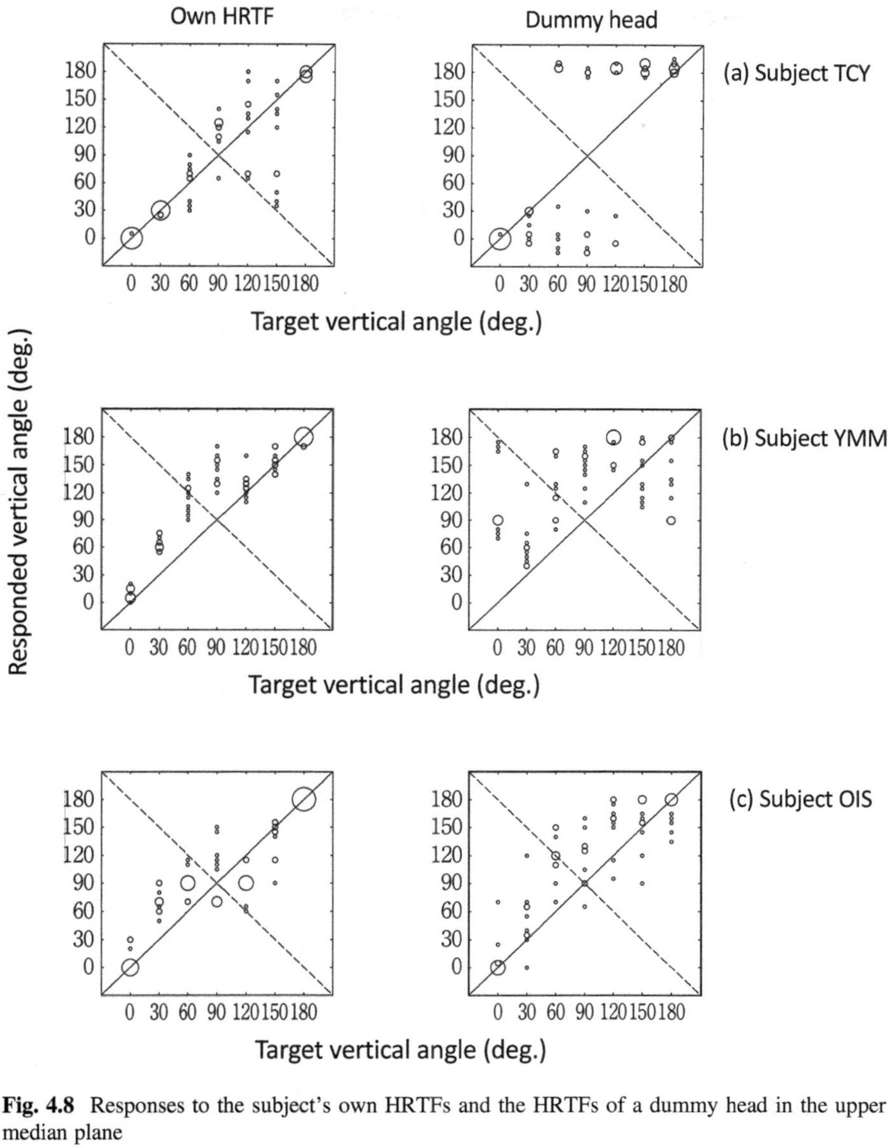

Fig. 4.8 Responses to the subject's own HRTFs and the HRTFs of a dummy head in the upper median plane

For 60° to 120°, the error of the dummy head is approximately twice those of the subject's own HRTFs and the actual sound sources. For the mean vertical localization error averaged over all directions (h), the error of the dummy head is approximately twice those of the subjects' own HRTFs and the actual sound sources.

The inside-of-head localization (lateralization) rates are shown in Table 4.12. For the subjects' own HRTFs, inside-of-head localization did not occur for any subjects or target vertical direction. For the HRTFs of the dummy head, however, subject

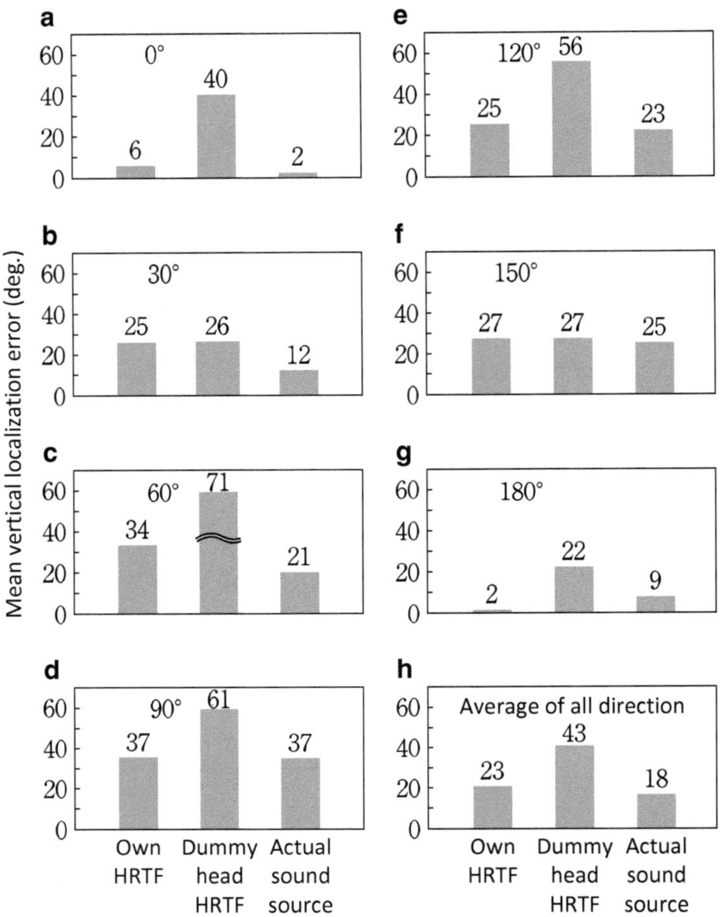

Fig. 4.9 Mean vertical localization error for the subject's own HRTFs, the HRTFs of a dummy head, and actual sound sources

TCY perceived a sound image in the head at the target vertical angles of 60°, 90°, and 120°. Subject YMM perceived a sound image in the head at all target vertical angles except 30°.

The front-back error rates of the median plane localization tests using eight kinds of dummy heads are shown in Fig. 4.10 (Møller et al. 1999). Whereas the front-back error rate for the actual sound sources was 16.0%, the error rates for the dummy heads were 37.3 to 50.2%. There existed significant differences between the actual sound source and each dummy head ($p < 0.01$).

As such, the HRTFs of dummy heads, which were developed from the representative shape of the head and pinna obtained from many subjects, do not provide accurate sound image localization for front and rear directions.

Table 4.12 Inside-of-head localization rate for the subject's own HRTFs and the HRTFs of a dummy head (%)

Subject	HRTF	Target vertical angle (deg.)						
		0	30	60	90	120	150	180
TCY	Own	0	0	0	0	0	0	0
	Dummy head	0	0	20	20	10	0	0
YMM	Own	0	0	0	0	0	0	0
	Dummy head	50	0	20	20	30	30	40
OIS	Own	0	0	0	0	0	0	0
	Dummy head	0	0	0	0	0	0	0

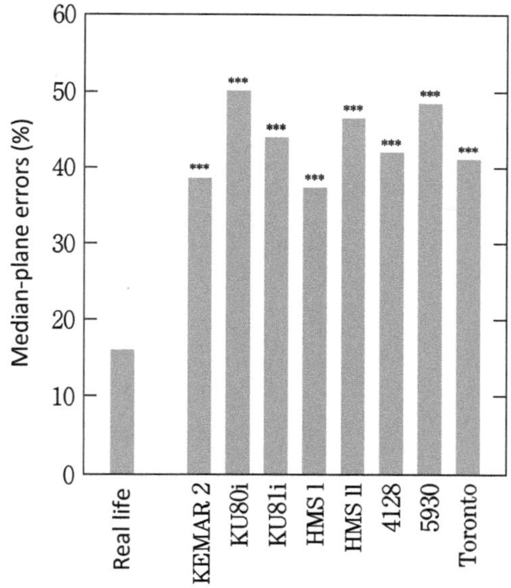

Fig. 4.10 Front-back error rate for actual sound sources and for the HRTFs of eight dummy heads (%) ***: $p < 0.001$. (Møller et al. 1999)

4.3.2 *Sound Image Localization Using Robust HRTF Sets*

Another approach to the standardization of HRTFs is to prepare robust HRTF sets, one of which allows most, but not all, of the listeners to perceive the three-dimensional acoustical sensation.

Until research on the generation of the listener's individual HRTFs discussed in Sect. 4.4 becomes practical, there is assumed to be no other choice but to use robust HRTF sets to develop the acoustic VR system. However, there remains a problem in that it is not possible to estimate beforehand whether an individual listener will be provided with accurate sound localization by such HRTF sets.

1. **Sound image localization using typical subjects**

Sound image localization tests through headphones were performed for 20 subjects using the HRTFs of 30 donors, including the subjects themselves (Møller et al. 1996). The target directions numbered 14, including seven directions in the median plane.

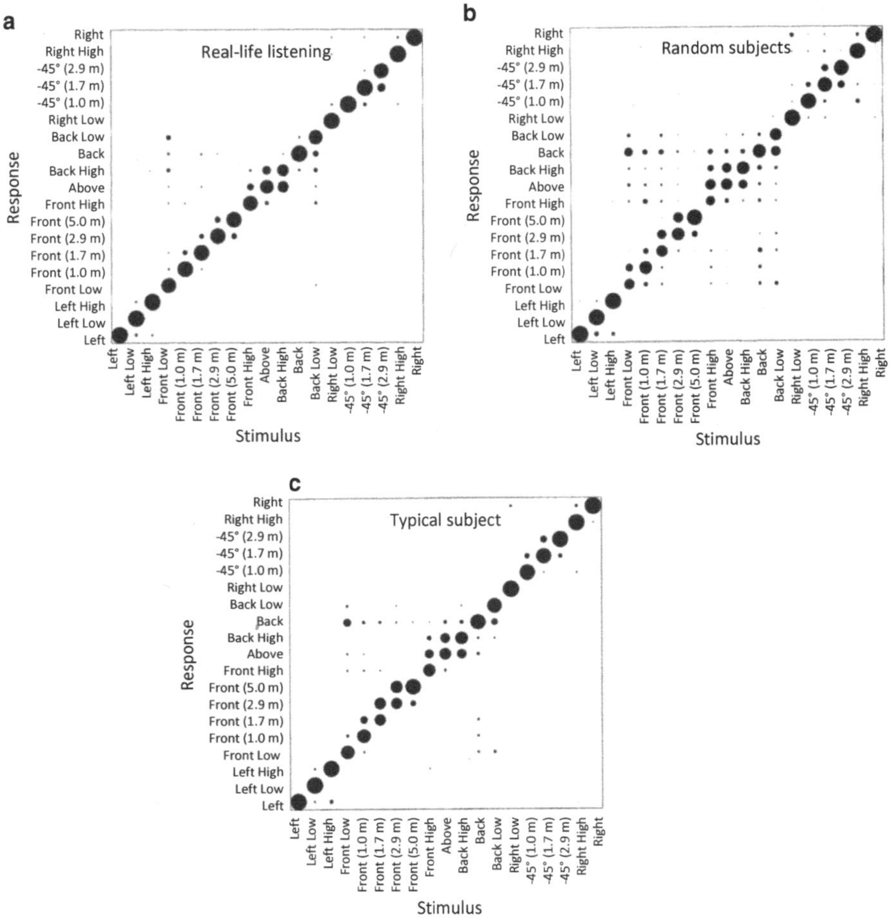

Fig. 4.11 Responses to actual sound sources, the HRTFs of random subjects, and the HRTFs of typical subjects (Møller et al. 1996). The values in the parentheses indicate the distance of a sound source. The distance of a sound source without parenthesis was 1.0 m

Figure 4.11(a) shows the responses to the actual sound sources. Most of the responses distributed around the direction of the actual sound source, while the responses were distributed widely for the target direction of the lower median plane. (See Appendix A.1 for details on the sound image localization accuracy for the actual sound source.)

Figure 4.11(b) shows all the responses to the HRTFs of the 30 donors (hereinafter the random subjects). Frequent front-back errors were observed in the median plane.

Figure 4.11(c) shows the responses to the HRTFs of the specific donor (hereinafter, the typical subjects) among the 30 donors, for which the mean localization error of the 20 subjects was the smallest. Localization errors were smaller than those shown in Fig. 4.11(b), although the front-back errors remained.

Table 4.13 Front-back error rate for actual sound sources, the HRTFs of random subjects, and the HRTFs of a typical subject (%). (Møller et al. 1996)

Real life	Random subjects	Typical subject
15.5	36.3	21.2

Furthermore, the front-back error rates are shown in Table 4.13. The error rate was 15.5% for the actual sound sources and 36.3% for the random subjects. Amazingly, the error rate for the random subjects is approximately the same as or lower than that for the dummy heads shown in Fig. 4.10. In other words, the dummy heads cause front-back errors as frequently or more than the random subjects.

On the other hand, the front-back error rate for the typical subject was 21.2%. Significant differences ($p < 0.001$) were observed among the actual sound source, the random subjects, and the typical subject.

These results suggest that determining a typical subject may realize three-dimensional sound image control, the accuracy of which is better than that for the dummy heads or the randomly selected HRTFs, although the accuracy is worse than that for the subject's own HRTFs. However, it is not clear how many donors are necessary to select the effective typical subject. Moreover, the range in application of the typical subject is also unknown.

2. **Sound image localization using typical HRTFs (small-scale localization tests)**

Expanding the idea mentioned above, the robust HRTFs for each target direction, the mean localization error of which is the minimum among subjects, were obtained.

Sound image localization tests in the median plane were performed for five subjects using the HRTFs of nine donors, including the subjects themselves. Figure 4.12 shows the results. The responses were distributed not only around the diagonal line, but also widely on the two-dimensional plane, and front-back errors occurred frequently.

The HRTFs (hereinafter, the typical HRTFs) for which the mean localization error of among the five subjects becomes smallest for each direction were selected. The responses of the five subjects to the typical HRTFs are shown in Fig. 4.13. Most of the responses distributed around the diagonal line, although the variances of the responses were somewhat large. These typical HRTFs are assumed to be applicable to the five subjects.

Figure 4.14 shows the amplitude spectrum of the HRTFs of nine subjects. The gray mesh area indicates the distribution range of the amplitude of the nine subjects. The black solid line indicates the amplitude spectrum of the typical HRTF for each target direction. It is observed that the typical HRTFs have remarkable notches and peaks at the front and rear directions (0°, 30°, 150°, and 180°). On the other hand, the amplitude spectra were flat at the upper directions (60°, 90°, and 120°). These observations suggest that the typical HRTFs may have the characteristics of the emphasized directional dependence of the HRTFs in the median plane.

Moreover, for the front and rear directions, the notch frequencies of the typical HRTFs are lower than those of other HRTFs. This supports the hypothesis "the

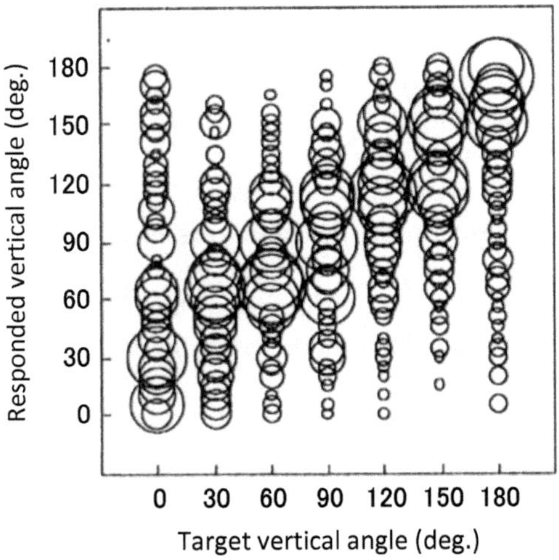

Fig. 4.12 Responses of five subjects to the HRTFs of nine donors

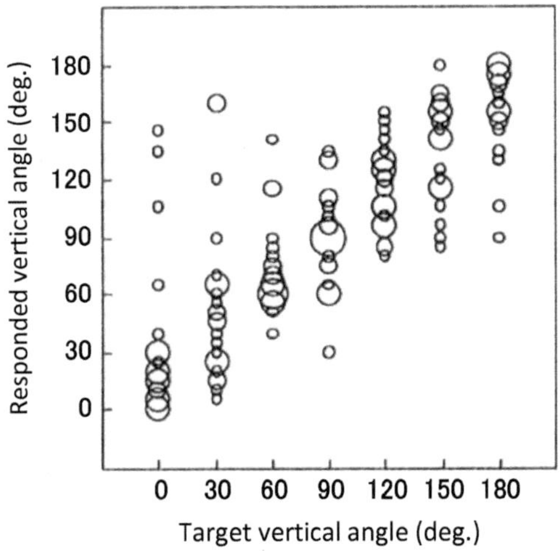

Fig. 4.13 Responses of five subjects to the typical HRTFs selected for each target vertical angle

HRTFs of large pinnae tend to provide good localization performance to listeners who have small pinnae" introduced in Sects. 2.2.2 and 3.8. However, it is not clear whether the typical HRTFs selected here provide accurate localization for other listeners.

3. **Sound image localization using typical HRTFs (large-scale localization tests)**

Furthermore, large-scale localization experiments searching for the typical HRTF sets at the seven directions in the upper median plane were performed using the HRTFs of 100 adult donors and 68 adult subjects who were not included among the 100 donors. The sound source was a wide-band white noise (200 Hz – 17 kHz).

The 68 subjects were divided into two groups. The typical HRTF sets were obtained by the following algorithm using the results of the localization tests for

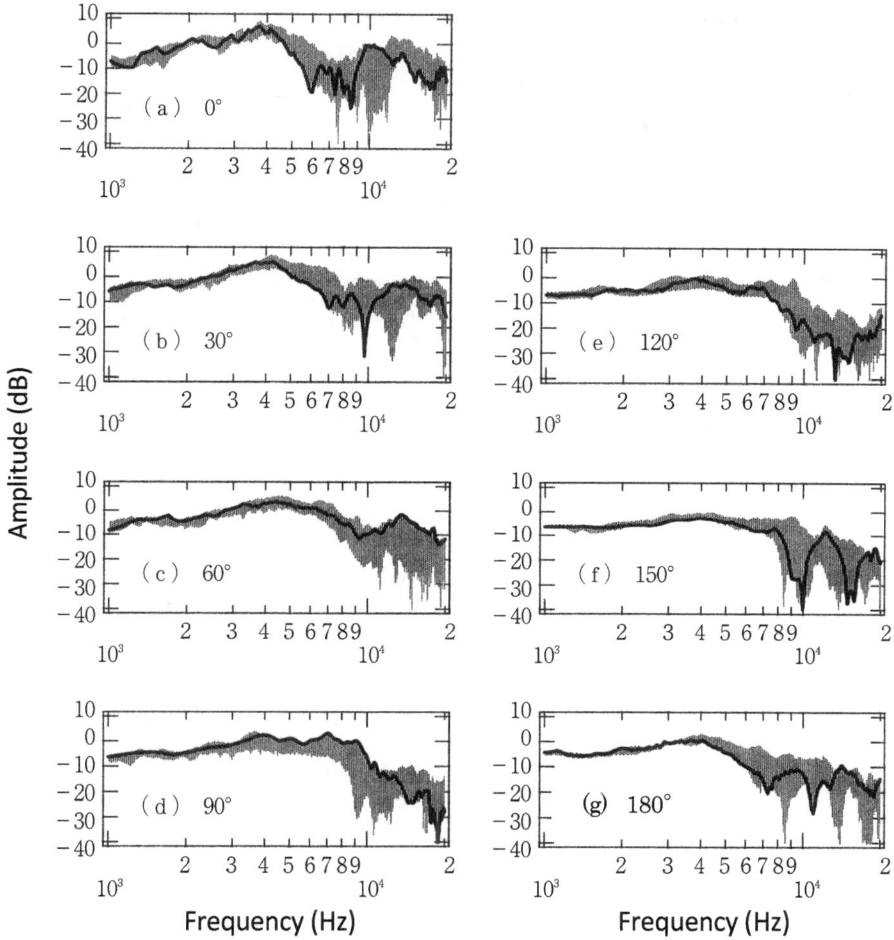

Fig. 4.14 Amplitude spectrum of the HRTFs of nine subjects. The gray mesh area indicates the distribution of the amplitude of nine subjects. The black solid line indicates the amplitude spectrum of the typical HRTF

group 1 (34 subjects). The results of the localization tests of group 2 were used for the validation of the obtained typical HRTF sets.

1. Select the 50 HRTFs for which the localization error averaged over 34 subjects were the smallest for each target direction.
2. Create all combinations, which choose the HRTFs of size n from among the 50 HRTFs (${}_{50}C_n$), where n denotes the number of typical HRTFs to be chosen.
3. Obtain the minimum localization error of each subject among the HRTFs for each combination of HRTFs.

4. Obtain the average of the minimum values over the subjects. Set the HRTFs for which the obtained average values were minimum as the typical HRTFs of size n.

Figure 4.15 shows the responses of group 1 to the obtained typical HRTFs for each size n. For n = 1, the responses were distributed widely. This means that none of the HRTFs of the 100 donors can provide accurate median plane localization for all of the 34 subjects. The distribution of the responses converged as n increased. The responses distributed around the diagonal line for n = 6.

Then, Fig. 4.16 shows the responses of group 2, the results of the localization tests of which were not used to choose the typical HRTFs, to the typical HRTFs of size n. As in group 1, the distribution of the responses converged as n increased. However, the distribution of the responses of group 2 was wider than that of group 1.

The mean localization errors of the typical HRTFs for each target vertical angle are shown in Fig. 4.17. The closed circle and the open circle denote the mean localization errors for group 1 and group 2, respectively. For comparison, the mean localization errors for the subject's own HRTFs and for the actual sound sources are also shown. Note that these errors were obtained from 49 subjects who were not included among the 68 subjects.

The mean localization errors for group 1 decreased as the number of typical HRTFs increased. The errors almost converged for n = 6 for each target vertical angle. The value of the convergence decreased as the target vertical angle increased. The values of the convergence were less than the mean localization errors of the subject's own HRTFs and the actual sound sources for all target vertical angles, except 0°.

The mean localization errors for group 2 also decreased as the number of typical HRTFs increased. The value of the convergence decreased as the target vertical angle increased. The values of the convergence were more than those of group 1. For $n \geq 6$, the values of the convergence were less than the mean localization errors of the subject's own HRTFs and the actual sound sources for all target vertical angles, except 0°. For the target vertical angle of 0°, the mean localization errors for the typical HRTFs were 25.4° and 23.8° for n = 6 and 10, respectively, whereas the errors were 14.5° and 8.8° for the subject's own HRTFs and the actual sound sources, respectively.

As described above, the mean localization errors of groups 1 and 2 for the typical HRTFs were less than those for the subject's own HRTFs and the actual sound sources. Here, I will discuss the reason for this. Figure 4.18 shows the responses to the subject's own HRTFs and the actual sound sources for 49 subjects, who were not included among the 68 subjects. These responses distributed as an inverted s-shaped curve centered over a diagonal line. Then, the responses tended to shift upward for the target vertical angles of 30°–150°.

On the other hand, the typical HRTFs were the HRTFs for which the localization errors were minimum. Therefore, the mean localization errors of the typical HRTFs may be less than those of the subject's own HRTFs and the actual sound sources.

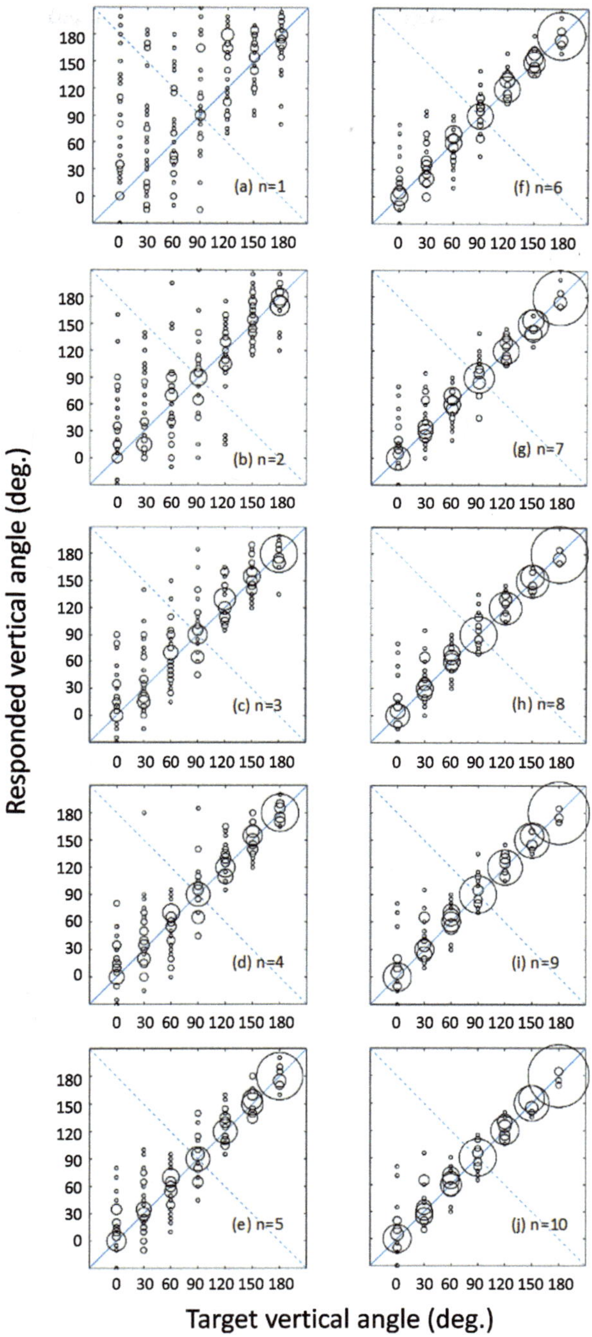

Fig. 4.15 Responses of subject group 1 to typical HRTFs. The term n denotes the number of typical HRTFs

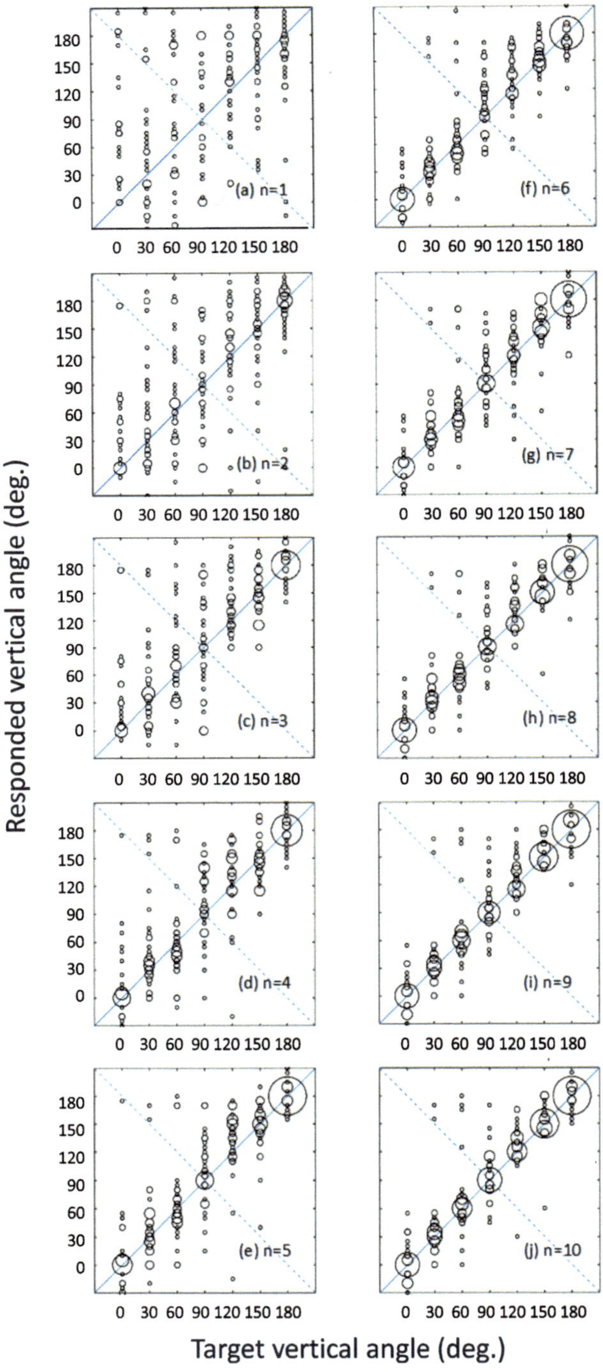

Fig. 4.16 Responses of subject group 2 (naive subjects) to typical HRTFs

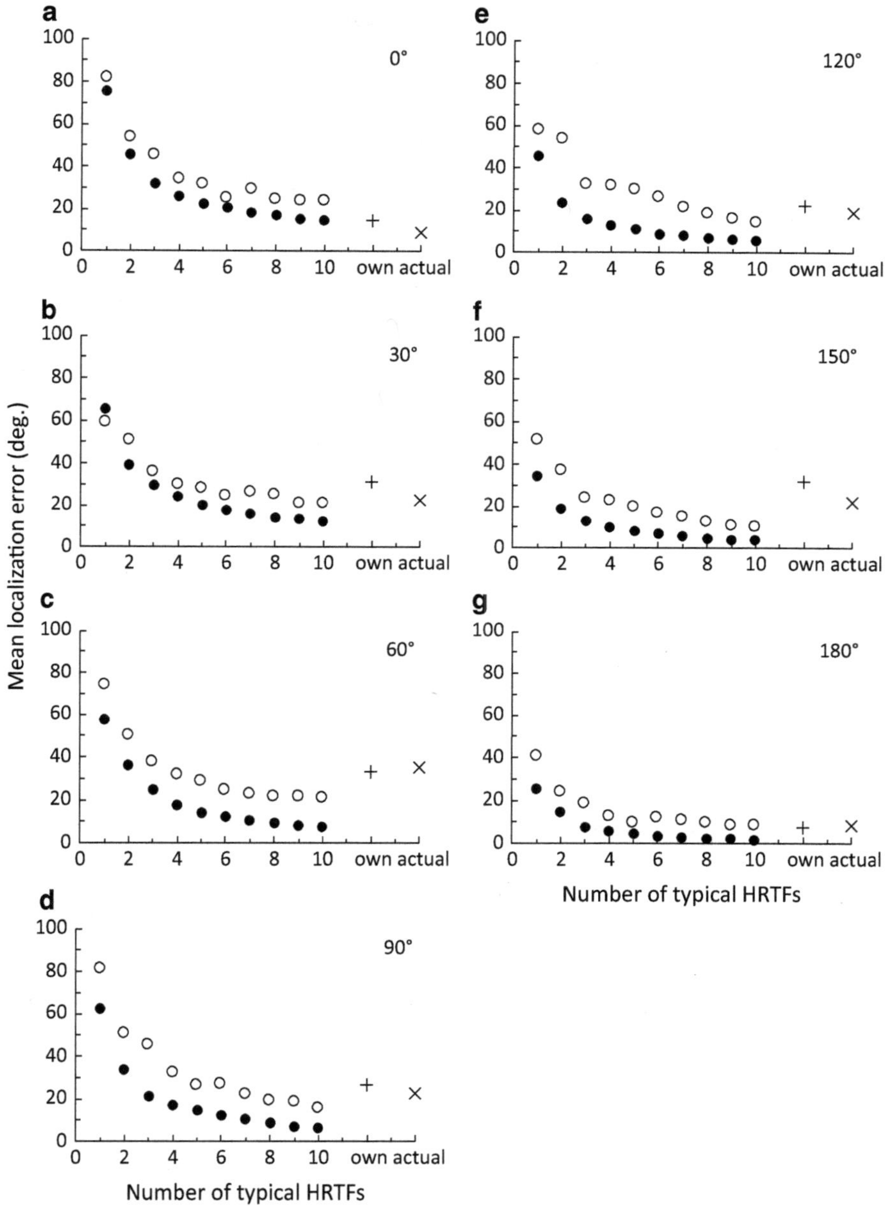

Fig. 4.17 Mean vertical localization error of typical HRTFs for subject group 1 (●) and subject group 2 (○). In the figure, + and × denote those of the subject's own HRTFs and of the actual sound sources, respectively

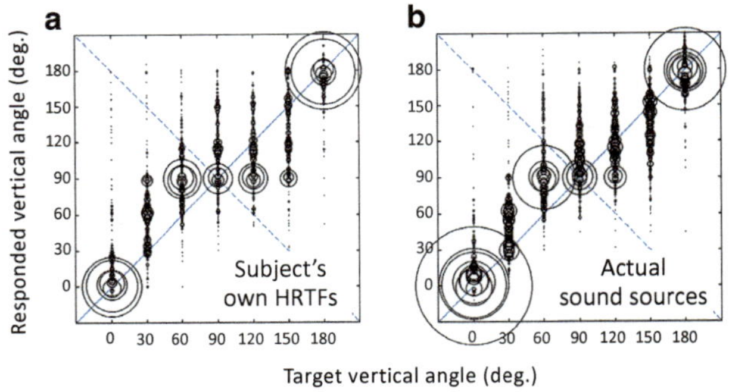

Fig. 4.18 Responses to (**a**) subject's own HRTFs and (**b**) actual sound sources in the upper median plane

4.4 Individualization of HRTFs

An authentic approach to realize an "acoustic VR "that can present three-dimensional acoustical sensation to everyone" described in the beginning of this chapter is to provide the HRTFs, which are exactly adapted to each listener. This approach is called the individualization (personalization) of HRTFs.

However, there exist two serious problems to realize the individualization of HRTFs. The first problem is that the characteristics of the amplitude spectrum of the HRTF are complicated. Therefore, it is necessary to clarify the essential information, in other words the cues for the perception of the sound image direction, and to discuss the individual difference in the cues. The second problem is that methods to estimate the concrete values of the cues of each listener have not succeeded.

In Chap. 2 and 3, the cues, by which a listener perceives the lateral angle and the vertical angle of a sound image, were described. Among these cues, the ITD and ILD are important cues to perceive the lateral angle of a sound image. Therefore, for the lateral angle, individualization of HRTFs should resolve only the second problem. In addition, some errors are assumed to be out of the question for the ITD and ILD estimation because the ITD and ILD change continuously and monotonically in accordance with the lateral angle of a sound source. The errors of individualization appear as a continuous quantity. For example, even if the sound image to be localized at the azimuth of 30° was perceived at 32° by a listener, serious problems would not occur in most applications.

On the other hand, for the vertical localization, cues for perception, spectral cues in other words, are under investigation, and, therefore, it is necessary to resolve both the first and second problems. Furthermore, vertical localization errors due to individual differences appear as discontinuously and fatal phenomena, such as front-back confusion, occur.

This section describes in detail the individualization of the HRTFs for vertical localization and then describes the individualization for lateral localization.

4.4.1 *Individualization of Amplitude Spectra of HRTFs*

The following approaches have been proposed as individualization methods of amplitude spectra of HRTFs:

1. Select the HRTF for which the pinna shape most closely resembles the listener's pinna shape from an HRTF database.
2. Expand or compress the amplitude spectrum of the reference HRTFs along the frequency axis.
3. Synthesize the HRTFs using the principal components based on the listener's pinna shape.
4. Estimate the spectral cues from the pinna shape of the listener and select the best-matching HRTFs from an HRTF database.
5. Generate the amplitude spectra of the individual HRTFs from the listener's pinna shape.
6. Select the suitable HRTFs from an HRTF database by listening tests.

Details of the above methods are described as follows.

1. **Select the HRTF for which pinna shape most closely resemble to the listener's pinna shape from an HRTF database**

This approach is based on the idea that if the pinna shape is similar, the HRTF is also similar (Zotkin et al. 2003). In this method, seven anthropometric parameters of a pinna of a listener are measured, and the HRTFs of a pinna for which the anthropometric parameters are closest to those of the listener are selected from the database (Fig. 4.19).

Sound image localization tests were performed using the selected HRTFs and the HRTFs of a dummy head in the anterior hemisphere. The results showed that the accuracy of localization for the selected HRTFs was slightly improved by 1.9° compared with the HRTFs of the dummy head.

The reason for the slight improvement in the localization accuracy is thought to be that the individual differences were calculated using all of the anthropometric parameters with same weighting. As described in Sect. 3.5, spectral peaks are generated by the modes of the depth direction and vertical direction of the pinna cavities. The notches are generated by multiple antinodes with different phases in the pinna. Therefore, the similarity of the pinna shape must be evaluated adding a great deal of weight to the anthropometric parameters related to the origins of the peaks and notches.

2. **Expand or compress the amplitude spectrum of the reference HRTFs along the frequency axis**

In this method, individual differences in the amplitude spectrum of directional transfer functions (DTFs), which is a direction-dependent component of a HRTF, are reduced by scaling along the frequency axis (Middlebrooks 1999a, b).

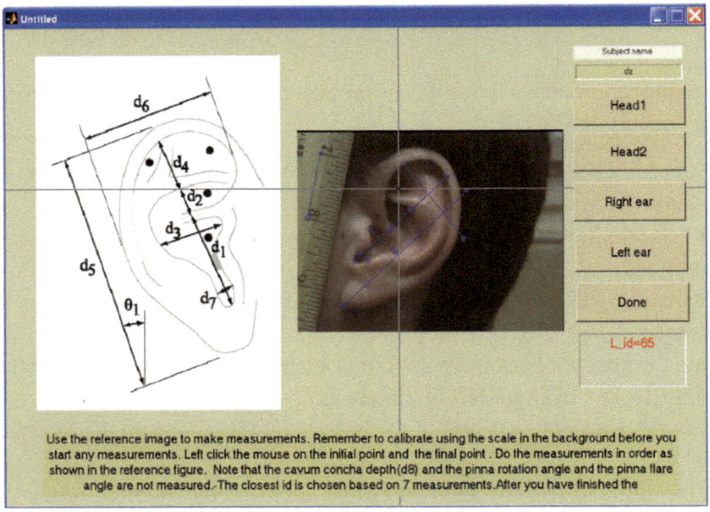

Fig. 4.19 Software that selects the HRTF of a pinna for which the anthropometric parameters are closest to those of the listener. (Zotkin et al. 2003)

Examples of frequency scaling are shown in Fig. 4.20. Sound image localization tests have shown that the quadrant error rate for DTFs, which were appropriately scaled from another's DTFs, was 14.7%. This value is approximately the same as the quadrant error for the listener's own DTFs (15.6%). Here, the quadrant error rate is defined as the error rate of over 90° in the front-back or up-down directions. In other words, the quadrant error rate is the sum of the front-back error rate and up-down error rate. However, one to three blocks of listening tests, each of which takes 20 min, were required to obtain the suitable scale factors for a listener.

3. **Synthesize the HRTFs using the principal components based on the listener's pinna shape**

A method that decomposes the amplitude spectrum of the HRTF into several principal components and synthesizes some of the components with weighting coefficients has been proposed (Kistler and Wightman 1992; Middlebrooks and Green 1992). The weighting coefficients depend on both the listener and the direction of a sound source and have been estimated based on the anthropometry of the listener's pinnae using multiple regression analysis (Reddy and Hegde 2015; Bomhardt et al. 2016) or using a deep neural network (Chun et al. 2017). However, the estimation of the weighting coefficients of an unknown listener has not been successful.

One study reported that 90% of amplitude spectra of the HRTF can be reproduced by the first five principal components (Kistler and Wightman 1992). On the other hand, another study reported that 12–18 principal components are required to reproduce prominent notches, which play an important role as spectral cues (Bomhardt et al. 2016), as shown in Fig. 4.21.

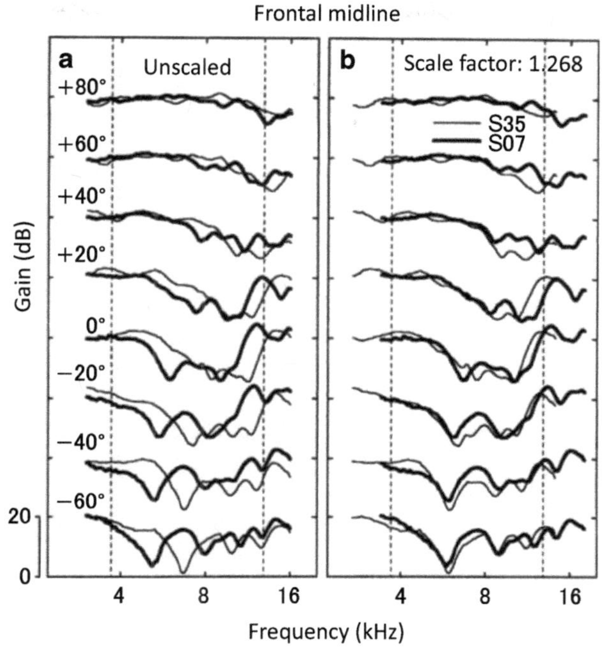

Fig. 4.20 Individualization of DTFs. Thin and thick lines represent measured DTFs of subjects S35 and S07, respectively. Unscaled DTFs are shown in (**a**). Scaled DTFs by 1.126 for subject S07 and 1/1.126 for subject S35 are shown in (**b**) (Middlebrooks 1999 a). sound sources in the upper median plane

4. **Estimate the spectral cues from the pinna shape of the listener and select the best-matching HRTFs from an HRTF database**

In this method, the frequencies of N1 and N2 are estimated from the pinna shape of the listener and the HRTFs having the frequencies of N1 and N2 that are closest to those are selected from the database (Iida et al. 2014).

A multiple regression analysis was performed using 56 ears (28 subjects) as objective variables of the frequencies of N1 and N2 of the front direction and as explanatory variables of ten pinna anthropometric parameters, as follows:

$$f(\mathrm{S})_{N1,N2} = a_1x_1 + a_2x_2 + \cdots + a_nx_n + \mathrm{b}(\mathrm{Hz}) \tag{4.2}$$

where S, a_i, b, and x_i indicate the subject, the multiple regression coefficient, a constant, and the dimension of the pinna anthropometry parameter, respectively.

The combination of parameters for which the correlation coefficient was the highest under the conditions whereby all of the p-values were less than 0.05 was adopted. As a result, six parameters (x_2, x_3, x_6, x_8, x_d, and x_a) were adopted for N1, and three parameters (x_6, x_8, and x_d) were adopted for N2 (Fig. 4.22). This means that the width, length, and depth of the pinna cavities and the tilt of the pinna correlated to N1, and the length and depth of the cavities correlated to N2. In these six parameters, x_6 and x_8 are regions related to the origin of the notches described in Sect. 3.4.3.

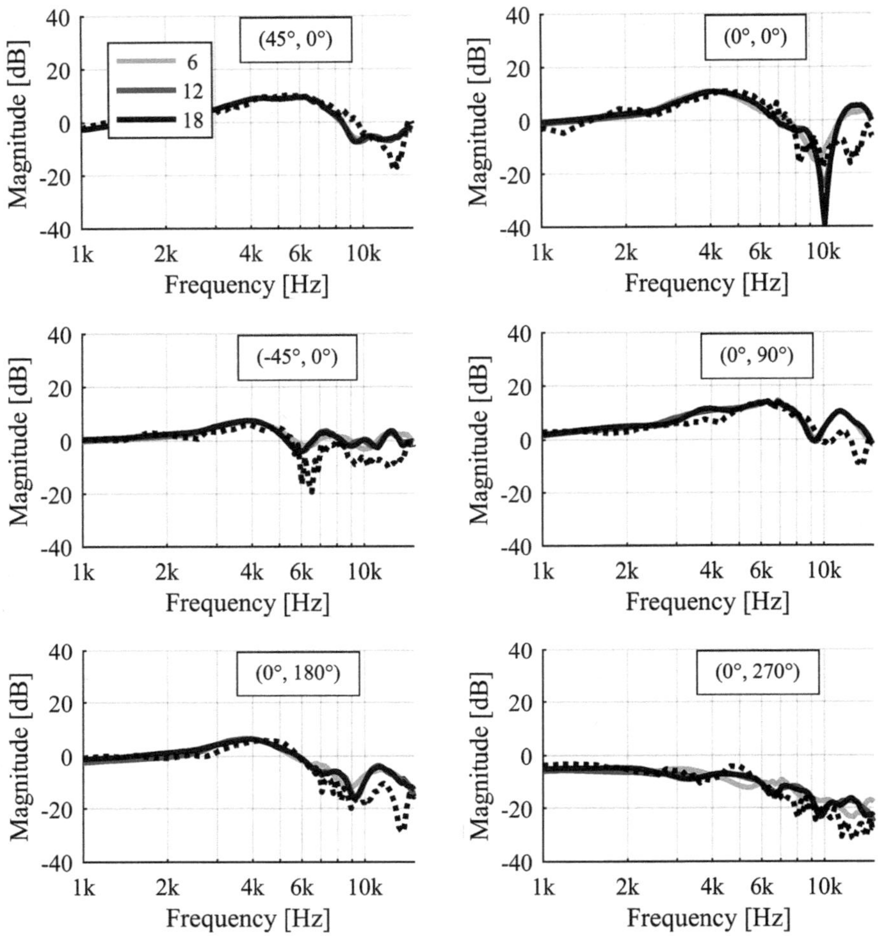

Fig. 4.21 Magnitude spectra of the original HRTFs (black dotted line) of data set 17 (chosen randomly) and the anthropometrically estimated HRTFs (from light to dark: 6 PCs, 12 PCs, and 18 PCs) are calibrated from real-valued PCs. (Bomhardt et al. 2016)

The multiple regression coefficient, p value, and 95% confidence interval are shown in Table 4.14. Figure 4.23 shows the relationship between the N1 and N2 frequencies estimated by the multiple regression model and those extracted from the measured HRTFs. The multiple correlation coefficients, r, of N1 and N2 are 0.81 and 0.82, respectively.

For the front direction, the frequencies of N1 and N2 of the four subjects, the frequencies of N1 and N2 and the pinna anthropometric parameters of which were not included in the multiple regression analysis, were estimated. Then, the HRTFs having frequencies of N1 and N2 that are closest to these frequencies are selected

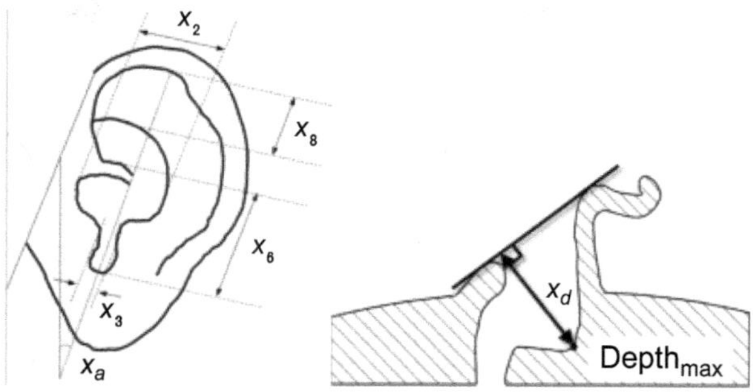

Fig. 4.22 Six anthropometric pinna parameters, which correlate to the frequencies of N1 and N2 for front direction. (Iida et al. 2014)

Table 4.14 Multiple regression coefficients, p-value, and 95% confidence intervals of N1 and N2 for the front direction. (Iida et al. 2014)

	Regression coefficient		p-value		95% confidence intervals			
					lower	upper	lower	upper
	N1	N2	N1	N2	N1		N2	
a_1								
a_2	116.9		1.6×10^{-2}		22.9	210.9		
a_3	−157.5		4.7×10^{-3}		−264.2	−50.8		
a_4								
a_5								
a_6	−183.4	−327.0	8.3×10^{-5}	2.9×10^{-7}	−269.1	−97.8	−438.0	−216.0
a_7								
a_8	−93.2	−245.0	2.3×10^{-3}	4.4×10^{-8}	−151.5	−34.9	−321.3	−168.6
a_d	−131.4	−172.8	4.0×10^{-3}	3.7×10^{-3}	−218.7	−44.2	−286.9	−58.7
a_a	−48.7		7.2×10^{-7}		−65.8	−31.6		
b	14906.4	23903.1	9.2×10^{-14}	2.0×10^{-22}	12019.9	17792.9	21079.9	26726.3

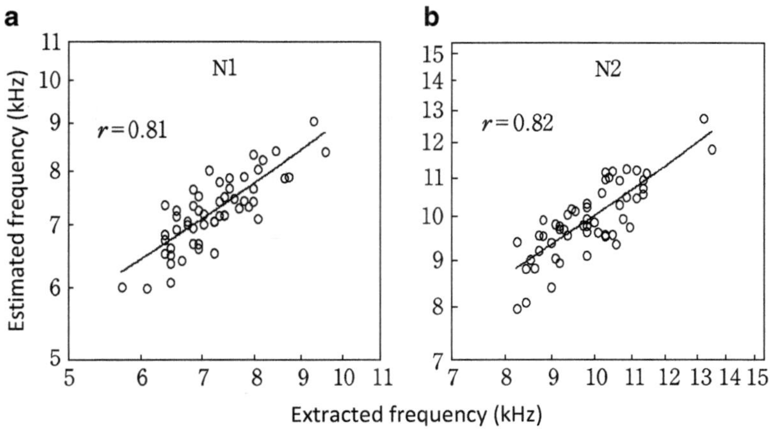

Fig. 4.23 Relationship between the frequencies extracted from the measured HRIR and the frequencies estimated from the listener's anthropometric parameters for 54 ears. (**a**) N1, (**b**) N2. In the figure, r denotes the correlation coefficient. (Iida et al. 2014)

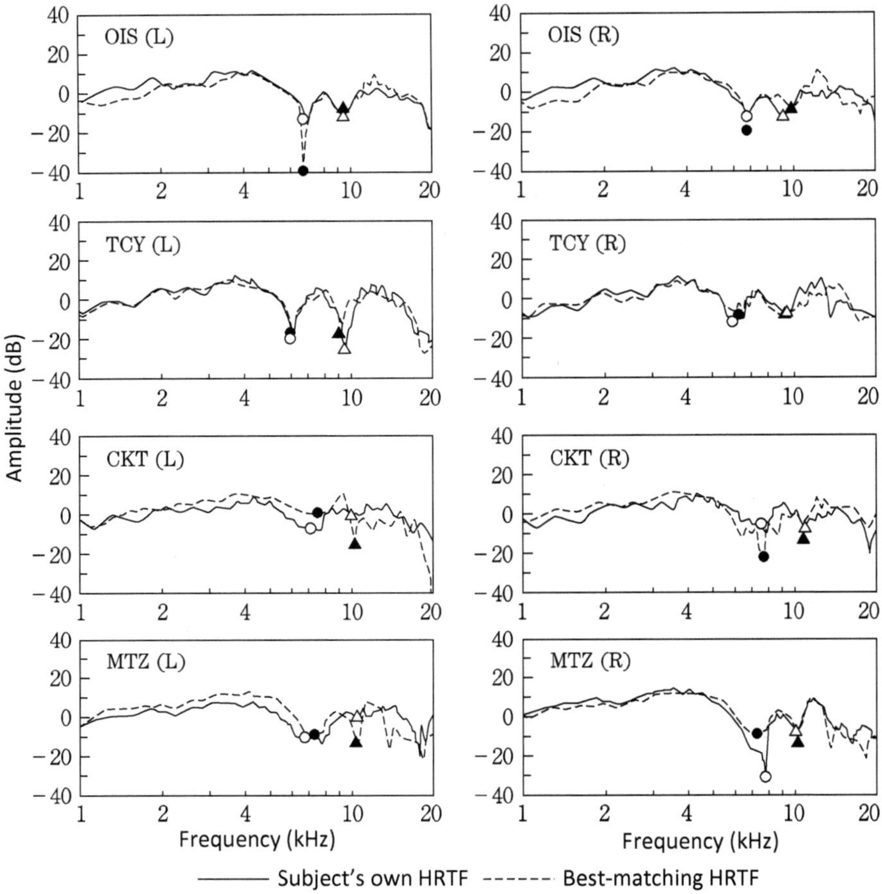

Fig. 4.24 Amplitude spectrum of best-matching HRTFs (broken line) and the subject's own HRTFs (solid line) for the front direction. (Iida et al. 2014). ●: N1 (best-matching), ○: N1 (subject's own), ▲: N2 (best-matching), △: N2 (subject's own)

from the database, as the best-matching HRTFs. The HRTFs for the other directions in the upper median plane were provided by the donor, for which the HRTF for the front direction was selected as the best match. Figure 4.24 shows the spectra of the measured HRTF and the best-matching HRTF of the four subjects.

The frequencies of N1 and N2 of the best-matching HRTFs (closed circles and triangles) and those of the measured HRTFs of the subjects (open circles and triangles) are close to each other, and the spectra of the best-matching HRTFs (dotted line) and subjects' measured HRTFs (solid line) were similar in almost all ears.

Furthermore, sound image localization tests for the seven directions in the upper median plane (interval of 30° from front to back) using the best-matching HRTFs were performed. The results (Fig. 4.25) indicate that the best-matching HRTFs

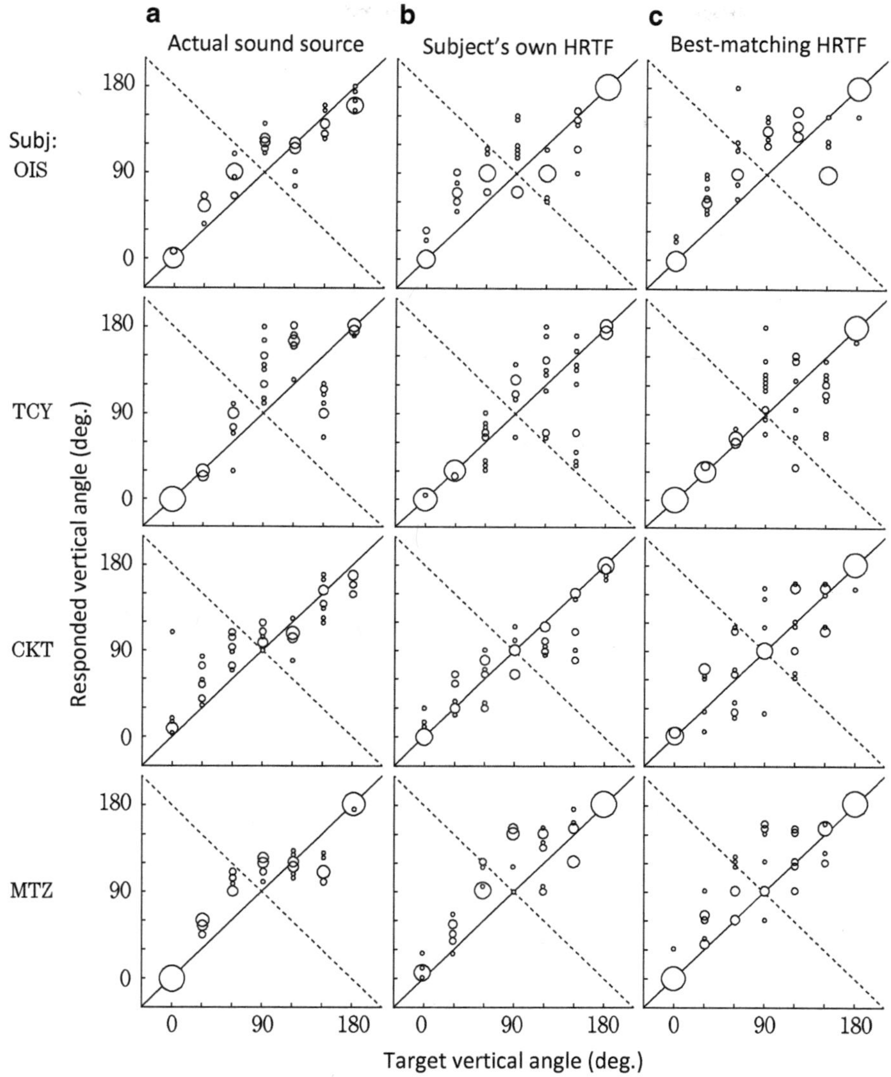

Fig. 4.25 Responses to (**a**) the actual sound sources, (**b**) the subject's own HRTFs, and (**c**) the best-matching HRTFs. (Iida et al. 2014)

provided approximately the same performance with respect to the perception of vertical angle as the subjects' own HRTFs for the target vertical angle of 0° (front), for which the N1 and N2 frequencies were estimated. For 180° (rear), the best-matching HRTFs provided approximately the same performance as for the target vertical angle of 0°. For the other five upper target directions, the performance of the localization for some of the subjects decreased compared to the subject's own HRTFs, and so there is room for improvement.

It has also been reported that the N1 frequency was estimated using eight pinna anthropometric parameters by a similar approach (Spagnol and Avanzini 2015). The report indicates that the distance from the entrance of the ear canal to the helix is the most important parameter.

On the other hand, the frequencies of peaks were also estimated from pinna shape. As shown in Table 4.15 and Fig. 4.26, the frequencies of P1 and P2 can be estimated from four and seven parameters, respectively.

From numerical computation by the FDTD method (Mokhtari et al. 2015; Mokhtari et al. 2016), the following relationships between pinna shapes and peak frequencies have been reported:

$$F_{p1} = 6461 - 758d_{B2-L4} - 439d_{B2-L5} \tag{4.3}$$

$$F_{p2} = 12441 - 1647d_{1-17} \tag{4.4}$$

$$F_{p3} = 15631 - 3298d_{4-12(vert)} \tag{4.5}$$

where F_{P1}, F_{P2}, and F_{P3} indicate the frequencies of P1, P2, and P3, respectively.

Here, $d_{B2\text{-}L4}$ indicates the distance from the bottom of the cavity of the concha to the outside of the antitragus, and $dB_{2\text{-}L5}$ indicates the distance from the bottom of the cavity of the concha to the side edge of the antihelixes. In addition, d_{1-17} indicates the distance from the entrance of the ear canal to the helix, and $d_{4-12\ (vert)}$ is the vertical length from the bottom of the cavity of the concha to the walls in front of the cymba concha. Figures 4.27 and 4.28 show the anthropometric parameters of the pinna. These parameters coincide with the explanatory parameters shown in Table 4.15. The multiple regression coefficients for the frequencies of P1, P2, and P3 were 0.84, 0.79, and 0.82, respectively. Therefore, good estimation of the P1, P2, and P3 frequencies is expected.

Table 4.15 Multiple regression coefficients, p-values, and 95% confidence intervals of P1 and P2 for the front direction

	Regression coefficient		p-value		95% confidence interval			
					lower	upper	lower	upper
	P1	P2	P1	P2	P1		P2	
a_1								
a_2		175.5		4.6×10^{-3}			57.0	294.0
a_3	−31.0	−145.8	2.2×10^{-2}	3.8×10^{-2}	−57.5	−4.6	−282.8	−8.7
a_4	−21.5		1.2×10^{-2}		−38.0	−5.0		
a_5	−19.5	51.0	1.7×10^{-4}	4.0×10^{-2}	−29.1	−9.8	2.5	99.5
a_6		−296.3		1.1×10^{-5}			−417.4	−175.2
a_7								
a_8		−203.6		2.5×10^{-5}			−291.2	−116.1
a_d	−35.3	−159.1	7.9×10^{-9}	5.6×10^{-3}	−60.9	−9.7	−269.1	−49.0
a_a		−42.5		2.8×10^{-4}			−64.3	−20.7
b	6627.4	16252.7	2.0×10^{-20}	9.9×10^{-11}	5762.8	7491.9	12322.6	20182.7

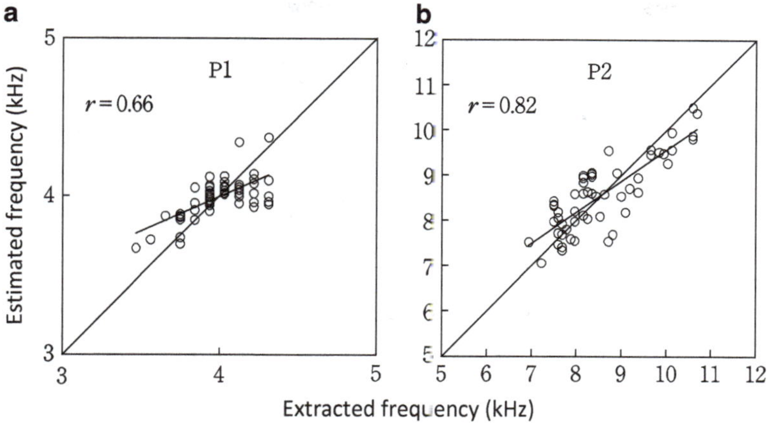

Fig. 4.26 Relationship between the frequencies extracted from the measured HRIR and the frequencies estimated based on the listener's anthropometric parameters. (**a**) P1, (**b**) P2. In the figure, r denotes the correlation coefficient

Fig. 4.27 Practical measurements of concha depth and aperture for F_{P1} estimation. F_{P1}-related concha depths $d_{B2\text{-}L4}$ and $d_{B2\text{-}L5}$. (Lateral landmarks L4 and L5 are indicated by small circles). (Mokhtari et al. 2015)

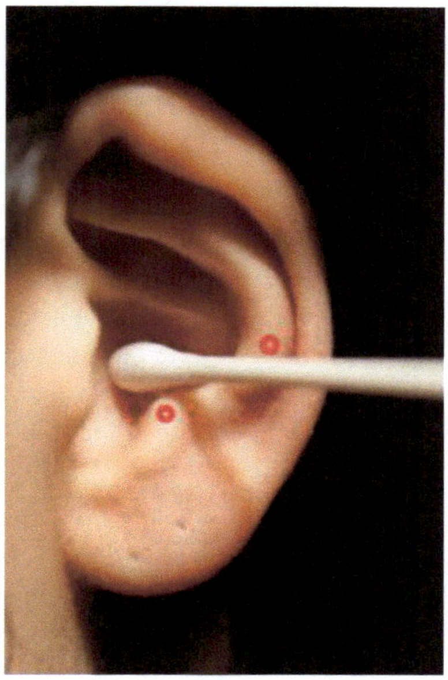

5. **Generate the amplitude spectra of the individual HRTFs from the listener's pinna shape**

This method generates the amplitude of the HRTFs of an unknown listener based on his/her pinna anthropometry and the partial regression coefficients, which were obtained beforehand by multiple regression analyses using several ears, as objective variables of

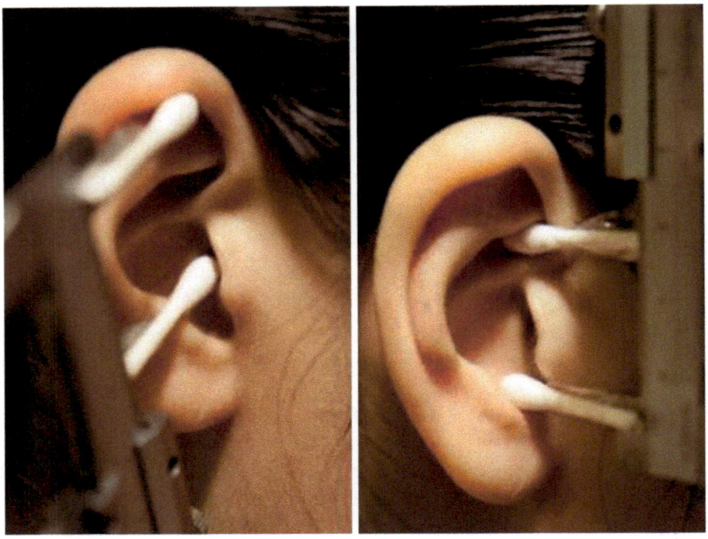

Fig. 4.28 Practical measurements for F_{P2} and F_{P3} estimation. Left panel: measurement of the distance d_{1-17} between the ear-canal entrance and the helix rim. Right panel: measurement of the vertical distance $d_{4-12(vert)}$ between the concha floor and the cymba anterior wall. (Mokhtari et al. 2016)

the amplitude of the measured HRTFs at each discretized frequency, and as explanatory variables of twelve pinnae anthropometry, as shown in Fig. 4.29 (Iida et al. 2019).

The results of the generation of the amplitude spectra of naive ears based on their pinna anthropometry and the partial regression coefficients, which were obtained by multiple regression analyses using 48 ears, are shown in Fig. 4.30. The generated amplitude spectra had similar prominent notches and peaks to those of the measured HRTFs, although some of the notches were shallow compared with the measured notches. The spectral distortion and the correlation coefficient between the amplitude spectra of the generated HRTFs and those of the measured HRTFs ranged from 4.6 to 6.2 dB and from 0.75 to 0.91, respectively. These results suggest that the outline of the amplitude spectrum of the generated HRTFs was similar to that of the measured early HRTFs, even though some absolute differences existed between them.

6. **Select suitable HRTFs from an HRTF database by listening tests**

A method by which HRTFs are selected by listening tests has also been proposed. Since the time required for the listening tests increases as the size of the database increases, methods to shorten the time for listening tests have been proposed.

One of the methods is a two-step choice method (Seeber and Fastl 2003). At the first step, a group of HRTFs is selected, and the best HRTF is selected from the group at the second step. It has been reported that this method takes approximately 10 min to select the HRTFs adapted to the listener.

Similarly, a tournament-style based method has also been proposed (Iwaya 2006). It has been reported that the time required for selecting HRTFs that adapt

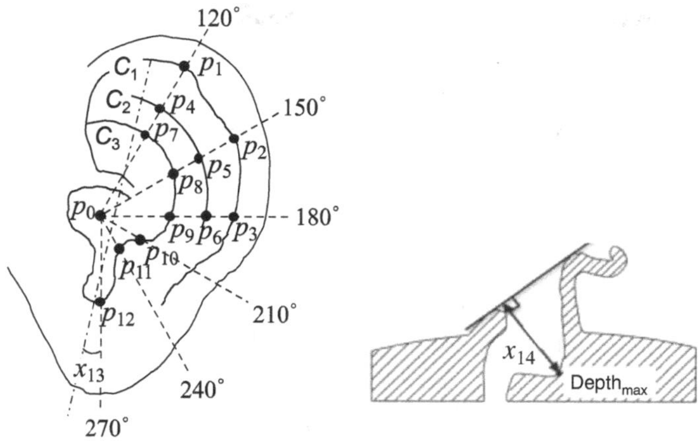

Fig. 4.29 Anthropometric parameters of the pinna (Iida et al. 2019). In the figure, x_1 through x_{12} were the lengths from p_0 to p_1 through p_{12}. In addition, x_{13} is the tilt of the pinna, and x_{14} is the depth of the concha cavity

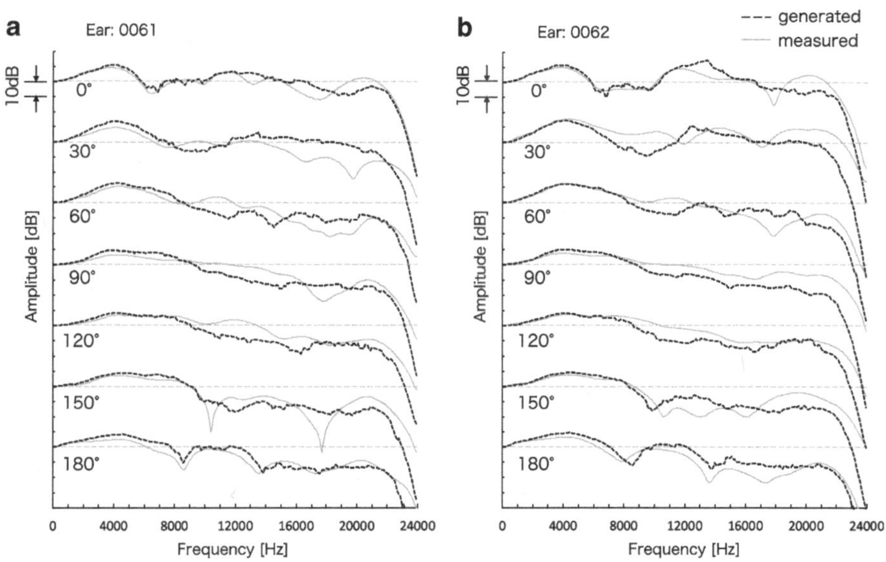

Fig. 4.30 Examples of the amplitude spectra of the generated HRTFs (broken line) and the measured early HRTFs (solid line) of two naive ears in the median plane. (**a**) Ear 0061 and (**b**) Ear 0062. (Iida et al. 2019)

to the listeners from among 32 kinds of HRTFs is shortened from 2 hours to 15 minutes by using this method. Figure 4.31 shows the results of sound image localization tests for the HRTFs selected from among 32 kinds of HRTFs by this method. The selected HRTFs provide approximately the same localization as the subject's own HRTFs, except when the target direction is to the rear. There exist

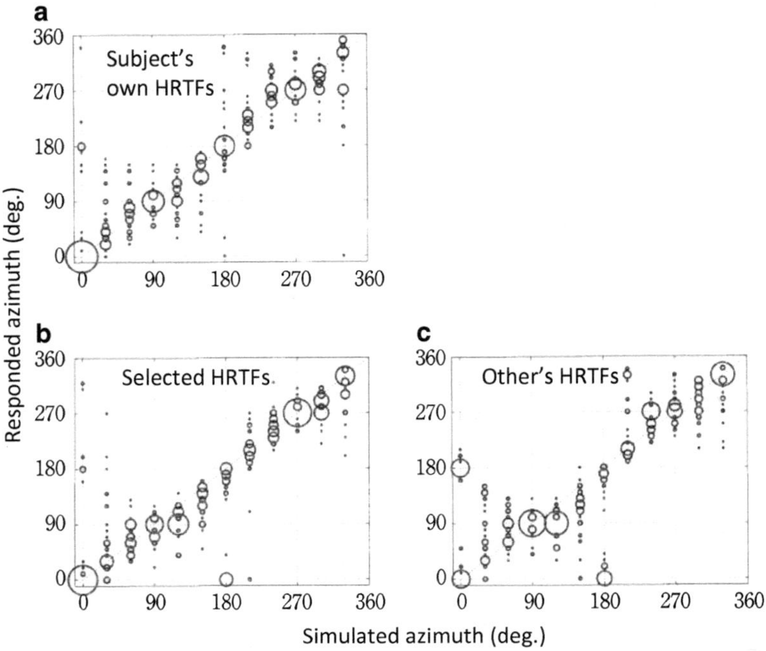

Fig. 4.31 Responses to (**a**) the subject's own HRTFs, (**b**) selected HRTFs, and (**c**) other's HRTFs in the horizontal plane. (Iwaya 2006)

errors to localize a sound image at the front when the target direction is to the rear. This suggests that more HRTFs to be selected should be included in the database.

4.4.2 Measures for Physical Evaluation of Individual Differences of HRTFs

In order to evaluate how the HRTFs obtained by methods [1] through [6] described above are adapted to the listeners, physical measures of individual differences of the HRTFs are required.

In order to evaluate the differences between the two HRTFs, $HRTF_j$ and $HRTF_k$, the spectral distortion (SD) shown by the following equation was used:

$$SD = \sqrt{\frac{1}{N}\sum_{i=1}^{N}\left[20\log_{10}\frac{|HRTF_{\mathrm{j}}(f_i)|}{|HRTF_{\mathrm{k}}(f_i)|}\right]^2}\ (\mathrm{dB}) \tag{4.6}$$

where f_i indicates discretized frequency.

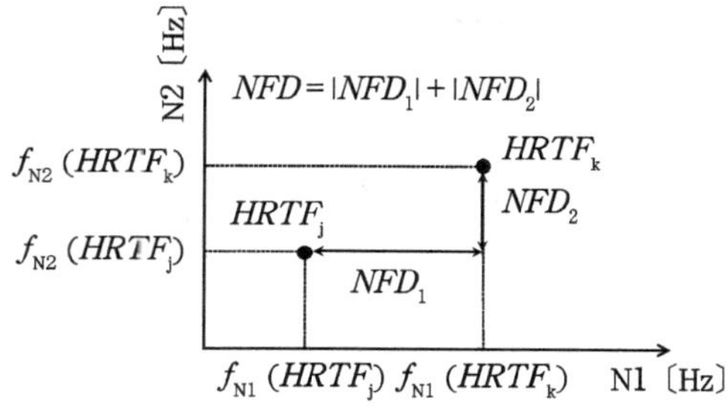

Fig. 4.32 Schematic diagram of NFD

The SD evaluates the differences in amplitude spectra of HRTFs for all frequency components uniformly. However, since there exist important frequency components (N1, N2, P1, and P2) and other components in the HRTFs, the individual differences should not be evaluated uniformly. It might be necessary to focus on the individual differences of the cues for perception of direction so as to evaluate the individual differences of HRTFs.

For vertical localization, the notch frequency distance (NFD) defined by Eqs. (4.7) through (4.9) has been proposed based on the idea that frequencies N1 and N2 are particularly important cues (Iida and Ishii 2011a). Figure 4.32 shows this concept.

$$NFD_1 = \log_2\left\{\frac{f_{N1}(HRTF_j)}{f_{N1}(HRTF_k)}\right\}(\text{octave}) \tag{4.7}$$

$$NFD_2 = \log_2\left\{\frac{f_{N2}(HRTF_j)}{f_{N2}(HRTF_k)}\right\}(\text{octave}) \tag{4.8}$$

$$NFD = |NFD_1| + |NFD_2|(\text{octave}) \tag{4.9}$$

where f_{N1} and f_{N2} indicate frequencies N1 and N2, respectively.

The following comparison was performed in order to examine the validity of the SD and the NFD. Figure 4.33 shows four HRTFs of the identical subject for the front direction, which was measured in the same anechoic chamber but in different years from 1999 to 2005. These HRTFs have similar outlines of the amplitude spectra, but the fine structures are different from each other. Even if there exist differences in fine structure, each of the four HRTFs provides accurate sound image localization to the subject (owner of the HRTFs).

Then, the SD and the NFD were calculated for all combinations, which chose two HRTFs from among the four HRTFs (${}_4C_2$). The results are shown in Tables 4.16 and 4.17. The SDs were 4.2 dB to 5.7 dB. These values were comparable with the SDs obtained using two HRTFs of other persons. In other words, SD was not able to distinguish the difference in HRTF measurements from the individual differences.

On the other hand, NFDs were 0.00 to 0.07 octaves. Differences in notch frequencies were within one sample (frequency resolution: 93.75 Hz) for N1 and three samples for N2. As described in Sect. 4.1, the JNDs of N1 and N2 were 0.1–0.2 octaves. The NFD for each combination of the four HRTFs was within the JNDs.

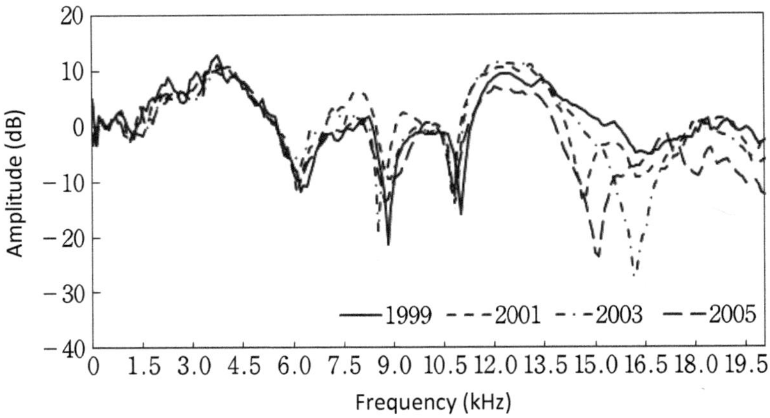

Fig. 4.33 Four HRTFs of a subject for the front direction, measured in the same anechoic chamber but in different years, from 1999 to 2005

Table 4.16 SDs for the four HRTFs shown in Fig. 4.32

	1999	2001	2003	2005
1999	–			
2001	4.2	–		
2003	5.1	4.3	–	
2005	5.7	4.8	5.6	–

Table 4.17 NFDs for the four HRTFs shown in Fig. 4.32

	1999	2001	2003	2005
1999	–			
2001	0.05	–		
2003	0.07	0.07	–	
2005	0.05	0.00	0.07	–

Therefore, the NFD is supposed to be able to explain the reason why each of the four HRTFs provided accurate sound image localization to the owner of the HRTFs.

4.4.3 Individualization of ITD

For estimation of the listener's individual ITD, numerous methods using the head and pinna shape of the listener have been studied.

A method to individualize ITDs that uses the spherical head model described in Chap. 2 (Fig. 2.6) has been proposed (Algazi et al. 2001). The ITDs of 25 subjects were measured, and a value of D, for which the ITD calculated by the following equation became closest to the measured ITD, was obtained for each subject:

$$\phi + \sin\phi = \frac{2c \times ITD}{D} \tag{4.10}$$

where ϕ, c, and D indicate the incidence azimuth (rad), speed of sound (m/s), and distance between both ears (diameter of the sphere) (m), respectively.

The RMSs of the estimation error of the ITD calculated by Eq. (4.10) were 22 to 47 μs for each sound source direction, and the average was 32 μs. The errors in azimuth were less than 5° for most of the sound source directions, and the maximum error was 12°. This indicates that accurate estimation of the individual ITD can be achieved when the listener's D is obtained.

A multiple regression analysis was performed as an objective variable of D and as explanatory variables of the width, length, and depth of the listener's head (Fig. 4.34):

$$\frac{D}{2} = w_1X_1 + w_2X_2 + w_3X_3 + \mathrm{b} \tag{4.11}$$

The obtained regression coefficients were $w_1 = 0.51$, $w_2 = 0.019$, $w_3 = 0.18$, and $b = 32$. However, the accuracy of the estimation of the ITD using this regression equation has not been shown.

On the other hand, the ITDs for twelve azimuth angles (30° steps) in the horizontal plane were estimated using ten head anthropometries (p_1 to p_7) shown in Fig. 4.35 and Table 4.18 (Ishii and Iida 2017). Table 4.19 shows simple correlation coefficients between ITDs and head anthropometries. Significant correlation was observed for the head anthropometries of p_1 and p_3 at several azimuth angles. These results agree with those of previous studies (Algazi et al. 2001; Watanabe et al. 2005).

For p_{4l} and p_{4r}, significant correlation was observed only in the case that the sound source was at the front half of the horizontal plane. For p_{5l} and p_{5r}, significant correlation was observed only in the case that the sound source was at the rear half of the horizontal plane. The correlation coefficients of the head anthropometry of the ipsilateral side of a sound source tended to be high compared with those of the contralateral side. For p_{6l}, p_{6r}, and p_7, significant correlation was observed for lateral directions.

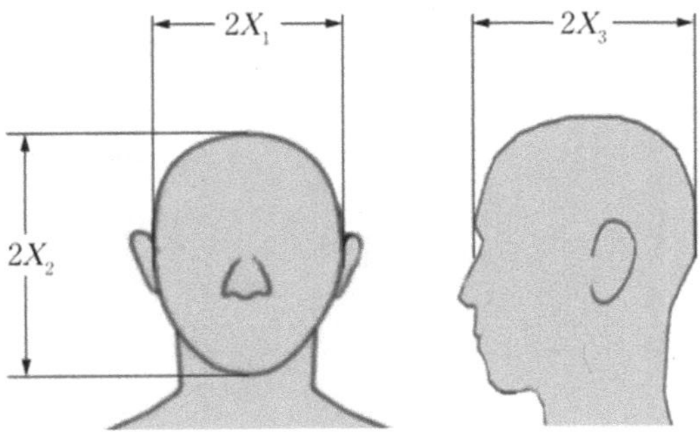

Fig. 4.34 Anthropometric parameters of the head for estimation of the distance between both ears, D. (Algazi et al. 2001)

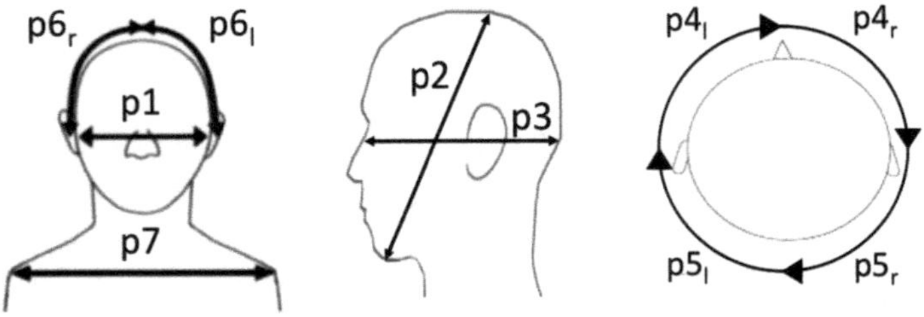

Fig. 4.35 Ten anthropometric parameters of the head. (Ishii and Iida 2017)

Table 4.18 Individual differences in ten anthropometric parameters of the head. (Ishii and Iida 2017)

	Ave.	Max	Min	Max - Min	Std.	RSD.
p_1	143	152	134	18	5.6	0.039
p_2	246	266	227	39	9.9	0.040
p_3	185	201	170	31	7.9	0.043
p_{4l}	152	167	136	31	7.4	0.049
p_{4r}	159	185	148	37	8.1	0.051
p_{5l}	144	163	130	33	8.1	0.056
p_{5r}	145	176	120	56	12.0	0.083
p_{6l}	195	217	176	41	8.2	0.042
p_{6r}	197	211	185	26	7.2	0.036
p_7	393	439	337	102	23.9	0.061

Table 4.19 Simple correlation coefficients between ITDs and head anthropometries. (Ishii and Iida 2017)

Azimuth (deg.)	p_1	p_2	p_3	p_{4l}	p_{4r}	p_{5l}	p_{5r}	p_{6l}	p_{6r}	p_7
0	−0.45	−0.20	−0.06	−0.40	−0.28	0.19	−0.09	−0.18	−0.17	−0.14
30	0.20	0.05	0.10	0.02	0.18	−0.04	0.01	−0.04	0.16	0.03
60	0.54	0.34	0.41	0.29	0.44	0.11	0.12	0.20	0.39	0.28
90	0.61	0.34	0.57	0.43	0.50	0.16	0.37	0.27	0.61	0.24
120	0.10	−0.03	0.38	0.04	0.02	0.28	0.38	0.09	0.22	−0.06
150	0.01	−0.02	−0.11	−0.10	−0.05	0.47	0.21	−0.29	0.14	−0.02
180	0.35	0.24	0.25	0.21	0.26	−0.17	0.11	0.14	0.14	0.05
210	0.00	0.20	0.05	0.12	0.07	−0.03	−0.02	0.25	0.06	−0.01
240	−0.23	−0.13	−0.23	−0.07	−0.16	−0.32	−0.09	−0.20	−0.26	−0.10
270	−0.65	−0.33	−0.46	−0.52	−0.58	−0.21	−0.34	−0.07	−0.64	−0.46
300	−0.54	−0.38	−0.42	−0.42	−0.63	0.15	−0.18	−0.01	−0.41	−0.51
330	−0.25	−0.16	−0.16	−0.19	−0.31	−0.09	0.02	0.08	−0.08	−0.02

Significant correlation tended to be observed for lateral directions. For 90° and 270°, the highest correlation was obtained for p_1 ($r = 0.61, -0.65$). On the other hand, for 30° and 210°, no head anthropometries showed significant correlation. These results indicate that simple linear regression analysis does not provide accurate estimation of the ITD at some azimuth angles.

Then, multiple regression analyses were performed as objective variables of the ITD for each of twelve azimuth angles in the horizontal plane and explanatory variables of ten head anthropometries, as follows:

$$ITD(s,\phi) = a_1(\phi)p_1(s) + a_2(\phi)p_2(s) + \cdots + a_7(\phi)p_7(s) + \mathrm{b} \quad (4.12)$$

where s and ϕ indicate the subject and the azimuth (°), respectively.

Table 4.20 shows multiple correlation coefficients (r), p-values of the entire regression model (p), the mean absolute residual (E), and partial regression coefficients. Multiple correlation coefficients ranged from 0.34 to 0.79, and significant correlation was observed for all twelve azimuth angles ($p < 0.05$). The mean absolute residual ranged from 9.6 to 24.8 μs.

The ITDs for the two females (A and B) and two males (C and D) in their twenties, who were not involved in the multiple regression analysis, were estimated using Eq. (4.12). The estimated ITD errors are shown in Table 4.21. Average of the estimated errors for each azimuth ranged from 8.1 to 26.5 μs and tended to be small for lateral directions (90° and 270°).

The estimated ITDs were converted into azimuth ϕ by Eq. (4.10). Table 4.22 shows the estimated azimuth errors. The azimuth errors averaged over the four naive subjects ranged from 1.1° to 3.3°.

4.4.4 Individualization of ILD

For estimation of the listener's individual ILD, a method using sine functions for each 1/3-octave band (Eq. 4.13) has been proposed (Watanabe et al. 2007)

$$\widehat{y_m} = \sum_{k=1}^{D} C_k \sin\frac{2\pi km}{M} \quad (4.13)$$

where the subscript m corresponds to the sound source direction $360(m^{-1})/M$ (°), $\widehat{y_m}$ indicates the ILDs, M indicates the number of sound source directions, C_k indicates the weight coefficient of the k-th sine function, and D indicates the order of the model, respectively. Here, C_k was obtained from the following multiple regression equation:

$$C_k(f_c) = a_{k1}x_1 + a_{k2}x_2 + \cdots + a_{k11}x_{11} + b_k \quad (4.14)$$

Table 4.20 Multiple regression coefficient, p-value, mean of absolute residual, regression coefficients, and constant. (Ishii and Iida 2017)

				Regression coefficient (x 10^{-3})										
Azimuth (deg.)	r	p	E(μs)	a_1	a_2	a_3	a_{4l}	a_{4r}	a_{5l}	a_{5r}	a_{6l}	a_{6r}	a_7	b
0	0.59	0.363	14.2	2.34	0.06	−0.72	1.11	−0.56	−0.48	−0.13	−0.21	−0.15	0.01	−127.24
30	0.34	0.975	17.9	−1.37	−0.34	−0.19	0.98	−0.33	0.24	0.18	0.43	−0.51	0.16	−197.13
60	0.63	0.210	15.0	−2.60	−0.27	−1.05	1.16	−0.02	0.20	0.55	−0.02	−0.62	0.02	−160.29
90	0.78	0.009	9.8	−0.92	0.08	−1.13	0.40	−0.30	0.21	−0.04	−0.18	−1.13	0.22	−278.76
120	0.65	0.165	20.8	1.50	1.23	−3.11	0.32	0.11	−0.15	−0.67	−0.83	−1.01	0.40	−295.37
150	0.71	0.057	12.2	−0.09	−0.27	−1.32	0.48	0.07	−1.09	−0.03	1.10	−0.41	−0.39	−163.68
180	0.50	0.672	16.0	−1.18	−0.61	−0.82	0.63	0.15	0.79	0.11	0.06	−0.03	0.28	185.10
210	0.38	0.949	21.1	1.59	−1.18	0.53	0.15	−0.90	−0.04	−0.41	−0.71	−0.15	0.33	458.68
240	0.43	0.874	24.8	0.79	0.15	0.63	−1.18	0.41	0.98	−0.17	0.53	0.71	−0.10	127.05
270	0.79	0.005	9.6	1.18	−0.14	−0.33	0.39	0.66	0.35	0.10	−0.59	1.07	0.04	320.99
300	0.75	0.022	18.4	2.00	−0.26	0.67	−0.72	1.08	−1.24	−0.37	−0.55	0.91	0.42	217.87
330	0.54	0.529	17.6	1.00	1.70	−1.24	−0.09	1.84	1.06	0.22	−1.46	−0.38	−0.67	142.87
Average	0.59	0.402	16.5											

Table 4.21 Estimation error in ITD (μs). (Ishii and Iida 2017)

Azimuth (deg.)	Subject A	B	C	D	Ave
0	21.4	24.4	35.2	15.0	24.0
30	21.4	40.9	15.1	17.0	23.6
60	20.1	53.1	11.1	12.8	24.3
90	20.0	4.1	23.2	12.2	14.9
120	49.4	2.2	15.6	38.9	26.5
150	33.5	25.4	7.6	32.5	24.8
180	0.2	44.2	20.2	18.9	20.9
210	4.2	50.0	19.8	17.0	22.7
240	26.5	9.4	0.1	20.0	14.0
270	2.5	7.6	9.0	13.3	8.1
300	5.4	17.7	29.1	7.8	15.0
330	11.8	8.0	4.9	22.0	11.7

Table 4.22 Estimation error in azimuth (deg.). (Ishii and Iida 2017)

Azimuth (deg.)	Subject A	B	C	D	Ave
0	–	–	–	–	–
30	2.4	4.9	1.6	1.8	2.7
60	2.6	7.1	1.4	1.4	3.1
90	–	0.7	4.3	2.2	2.4
120	6.7	0.3	2.0	4.3	3.3
150	4.0	2.7	0.9	3.2	2.7
180	–	–	–	–	–
210	0.5	7.1	2.2	1.9	2.9
240	3.1	1.3	0.0	2.3	1.7
270	0.0	1.4	1.7	2.4	1.4
300	0.7	2.2	3.4	0.8	1.8
330	1.3	0.7	0.5	1.9	1.1

where f_c, a_{ki}, x_i, and b_k indicate the center frequency of the 1/3-octave band, the multiple regression coefficient, the head shape parameter shown in Fig. 4.36, and a constant, respectively.

Figure 4.37 shows the ILDs for three 1/3-octave bands estimated by this method. The solid line, dashed line, and dotted line indicate measured ILDs, ILDs modeled by Eq. (4.14), and ILDs estimated from head shapes, respectively. The estimated ILDs were approximately the same as the measured ILDs.

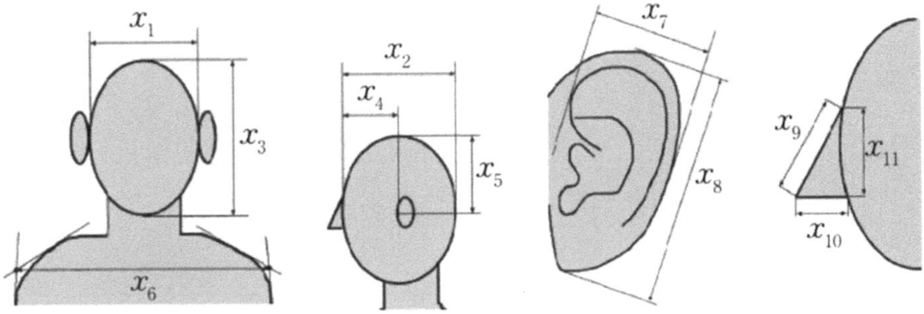

Fig. 4.36 Eleven anthropometric parameters of the head. (Watanabe et al. 2007)

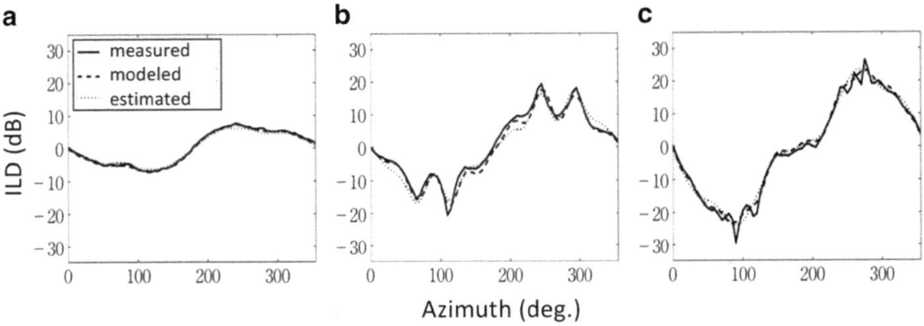

Fig. 4.37 ILD in the horizontal plane. Solid lines, broken lines, and dotted lines indicate measured ILD, ILD modeled by Eq. (4.14), and ILD estimated from the head anthropometry. (**a**) 1/3 oct. band center frequency of 500 Hz, (**b**) 2 kHz, (**c**) 5 kHz. (Watanabe et al. 2007)

4.4.5 Expected Future Development

In order to evolve acoustic VR into a universal system that can present three-dimensional acoustical sensation to everyone, the individualization method of HRTFs of an unknown listener, which do not require acoustical measurements, must be established.

As described above, several methods for HRTF individualization have been proposed. However, problems associated with practical use remain for each method. Here, I will summarize the problems and expected future developments for each method.

For the method using PCA, the amplitude spectrum of the HRTF is synthesized using principal components and weighting coefficients. The weighting coefficients depend on both the listener and the direction of a sound source. The weighting coefficients have been estimated based on the anthropometry of the listener's pinnae using multiple regression analysis or using a deep neural network. However, the estimation of the weighting coefficients of an unknown listener has not been successful. An accurate estimation method of individual listener's weighting coefficients must be established.

For the method, which estimates the prominent spectral peaks and notches, it is demonstrated that the minimum HRTF components, which provide approximately the same localization performance as the measured HRTFs, were the two lowest-frequency notches and the two lowest-frequency peaks. The frequency of the two notches and the two peaks were reported to be estimated based on the anthropometry of the listener's pinnae. However, estimation of the level of the notches and peaks has not been successful. A method that estimates the level and sharpness (band width) of the notches and peaks of an individual listener must be established.

The method, which estimates the amplitude level of the individual HRTFs at each discretized frequency based on the anthropometry of the pinnae and the partial regression coefficients, generated an outline of the amplitude spectrum of the individual HRTFs, which is similar to that of the measured early HRTFs. However, there existed some absolute difference between the HRTFs.

Numerical calculation of HRTFs has also been studied intensively. The boundary element method (BEM) has been used to calculate HRTFs in a number of studies (Katz 2001; Kahana and Nelson 2006; Kreuzer et al. 2009). The results of numerical calculations by the finite-difference time-domain (FDTD) method, which is much faster than the BEM, revealed that the fundamental spectral feature of the HRTF of an individual listener can be calculated from the baffled pinna (Takemoto et al. 2012). At present, however, neither the BEM nor the FDTD method is available for ordinary listeners because special equipment, e.g., a functional magnetic resonance imaging system, is required to digitize the complicated shape of the pinnae of an individual listener. Namely, at the moment, calculation of the HRTF is available only to a limited number of listeners. The development of an easy method by which to model the head and pinnae is necessary for the calculation of HRTFs for general listeners.

References

Algazi VR, Avendano C, Duda RO (2001) Estimation of spherical–head model from anthropometry. J Audio Eng Soc 49:472–479

Bernstein LR (2004) Sensitivity to interaural intensitive disparities:listeners' use of potential cues. J Acoust Soc Am 115:3156–3160

Bomhardt R, Braren H, Fels J (2016) Individualization of head-related transfer functions using principal component analysis and anthropometric dimensions. Proc of Meetings on Acoustics 29:050007

Burkhard MD, Sachs RM (1975) Anthropometric manikin for acoustic research. J Acoust Soc Am 58:214–222

Chun CJ, Moon JM, Lee GW, Kim NK, Kim HK (2017) Deep neural network based HRTF personalization using anthropometric measurements. Audio Eng Soc Convention 143:9860

Domnitz RH, Colburn HS (1977) Lateral position and interaural discrimination. J Acoust Soc Am 61:1586–1598

Hartmann WM, Constan ZA (2002) Interaural level differences and the level–meter model. J Acoust Soc Am 112:1037–1045
Hershkowitz RM, Durlach NI (1969) Interaural time and amplitude jnds for a 500- Hz tone. J Acoust Soc Am 46:1464–1467
Iida K, Ishii Y (2011a) 3D sound image control by individualized parametric head-related transfer functions in proc. Inter-Noise 2011: 428959, Osaka, Japan
Iida K, Ishii Y (2011b) Individualization of the head-related transfer functions in the basis of the spectral cues for sound localization. In: Suzuki Y, Brungart D, Iwaya Y, Iida K, Cabrera D, Kato H (eds) Principles and applications of spatial hearing. World Scientific Publishing, Singapore, pp 159–178
Iida K, Ishii Y, Nishioka S (2014) Personalization of head–related transfer functions in the median plane based on the anthropometry of the listener's pinnae. J Acoust Soc Am 136:317–333
Iida K, Shimazaki H, Oota M (2019) generation of the individual head-related transfer functions in the upper median plane based on the anthropometry of the listener's pinnae. Appl Acoust 155:280–285
Ishii Y, Iida K (2017) Personalization of interaural difference cues based on the anthropometry of the listener's head – estimation of interaural time difference –. Transaction of the Virtual Reality Society of Japan. 22: 405–412 (in Japanese)
Iwaya Y (2006) Individualization of head–related transfer functions with tournament– style listening test:listening with other's ears. Acoust Sci Tech 27:340–343
Kahana Y, Nelson PA (2006) Numerical modelling of the spatial acoustic response of the human pinna. J Sound Vibration 292:148–178
Katz BFG (2001) Boundary element method calculation of individual head–related transfer function. I. Rigid model calculation. J Acoust Soc Am 110:2440–2448
Kistler DJ, Wightman FL (1992) A model of head–related transfer functions based on principal components analysis and minimum–phase reconstruction. J Acoust Soc Am 91:1637–1647
Kreuzer W, Majdak P, Chen Z (2009) Fast multipole boundary element method to calculate head-related transfer functions for a wide frequency range. J Acoust Soc Am 126:1280–1290
Middlebrooks JC (1999a) Individual differences in external–ear transfer functions reduced by scaling in frequency. J Acoust Soc Am 106:1480–1492
Middlebrooks JC (1999b) Virtual localization improved by scaling nonindividualized external–ear transfer functions in frequency. J Acoust Soc Am 106:1493–1510
Middlebrooks JC, Green DM (1992) Observations on a principal components analysis of head–related transfer functions. J Acoust Soc Am 92:597–599
Mills AW (1958) On the minimum audible angle. J Acoust Soc Am 30:237–246
Mokhtari P, Takemoto H, Nishimura R, Kato H (2015) Frequency and amplitude estimation of the first peak of head–related transfer functions from individual pinna anthropometry. J Acoust Soc Am 137:690–701
Mokhtari P, Takemoto H, Nishimura R, Kato H (2016) Vertical normal modes of human ears: individual variation and frequency estimation from pinna anthropometry. J Acoust Soc Am 140:814–831
Møller H, Jensen CB, Hanmmershøi D, Sørensen MF (1996.5) Using a typical human subject for binaural recording. Audio Eng Soc Reprint 4157 (C–10)
Møller H, Hanmmershøi D, Jensen CB, Sørensen MF (1999) Evaluation of artificial heads in listening tests. J Audio Eng Soc 47:83–100
Reddy S, Hegde RM (2015) A joint sparsity and linear regression based method for customization of median plane. IEEE Asilomar, 785–789
Seeber BU, Fastl H (2003) Subjective selection of non–individual head–related transfer functions, Proceedings of the 2003 international conference on auditory display, Boston, MA, USA, July 6–9, 2003 ICAD03–(1–4)
Spagnol S, Avanzini F (2015) Frequency estimation of the first pinna notch in head–related transfer functions with a linear anthropometric model. Proc. of the 18th Int. conference on digital audio effects(DAFx-15), Trondheim, Norway, Nov 30 - Dec 3

Takemoto H, Mokhtari P, Kato H, Nishimura R, Iida K (2012) Mechanism for generating peaks and notches of head–related transfer functions in the median plane. J Acoust Soc Am 132:3832–3841

Watanabe K, Iwaya Y, Gyoba J, Suzuki Y, Takane S (2005) An investigation on the estimation of interaural time difference based on anthropometric parameters TVRSJ. 10-4 : 609–618. (in Japanese)

Watanabe K, Ozawa K, Iwaya Y, Suzuki Y, Aso K (2007) Estimation of interaural level difference based on anthropometry and its effect on sound localization. J Acoust Soc Am 122:2832–2841

Zotkin DN, Hwang J, Duraiswami R, Davis LS (2003) HRTF personalization using anthropometric measurements. IEEE Workshop on Applications of Signal Processing to Audio and Acoustics

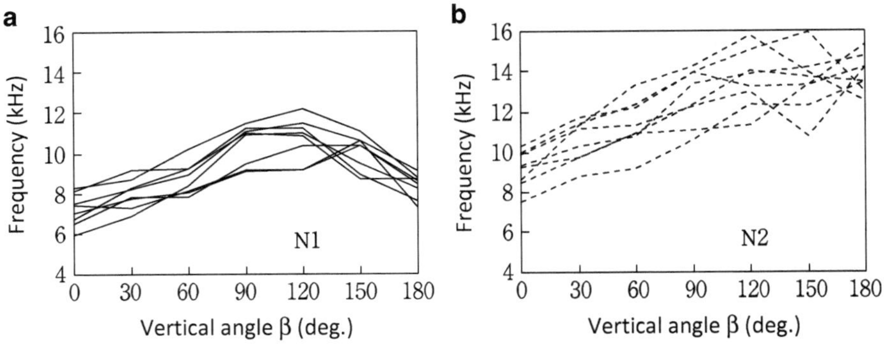

Fig. 5.1 Frequencies of N1 and N2 in the upper median plane for six subjects. (**a**) N1 and (**b**) N2. (Iida and Ishii 2011)

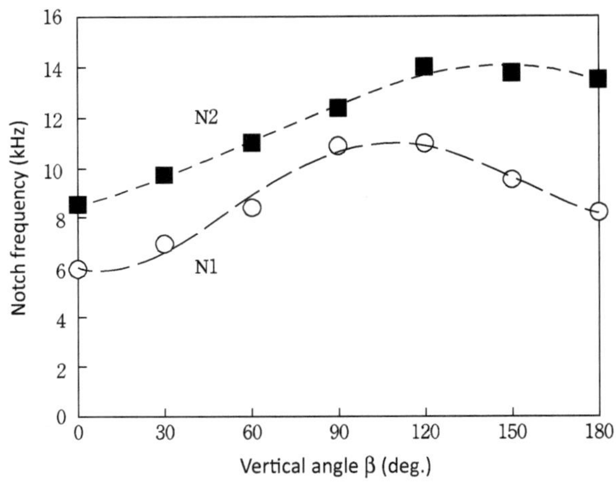

Fig. 5.2 Examples of the measured (open circle and closed square) and estimated (broken lines) N1 and N2 frequencies in the upper median plane. (Iida and Ishii 2011)

$$f_{N2}(\beta) = 1.310 \times 10^{-5} \times \beta^4 - 5.154 \times 10^{-3} \times \beta^3 + 5.020 \times 10^{-1} \times \beta^2 + 2.565 \times 10 \times \beta + f_{N2}(0) \quad (5.2)$$

Here, β indicates the vertical angle of a sound source, and $f_{N1}(0)$ and $f_{N2}(0)$ indicate the N1 and N2 frequencies of an individual listener for the front direction, respectively.

5.2 Similarity of Notches and Peaks Among Sagittal Planes

Focusing on the cues of perception of the vertical angle of a sound image described in Chap. 3, more efficient sound image control is expected for a three-dimensional direction. It has been observed that the amplitude spectrum of the HRTF did not vary greatly in the transverse section. It is also known that the directional bands (see

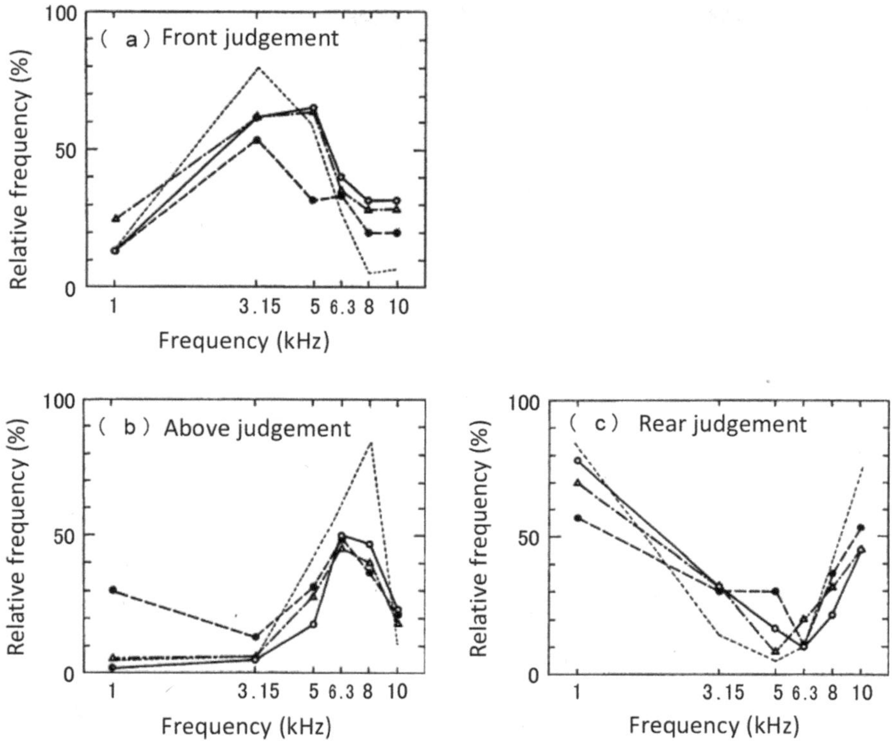

Fig. 5.3 Relative frequencies of (**a**) front, (**b**) above, and rear judgments. Open circles and open triangles indicate the sagittal plane, the lateral angles of which are 30° and 60°. Closed circles indicate a lateral angle of 90°. (Morimoto and Aokata 1984). Dotted lines indicate Blauert's data. (Blauert 1969/70)

Chap. 6) occur in either sagittal plane, the lateral angles of which are 0° (median plane), 30°, and 60°, as shown in Fig. 5.3 (Morimoto and Aokata 1984). These findings suggest that the spectral cue of the vertical angle is common among sagittal planes.

In order to verify this hypothesis from a physical viewpoint, the similarity of the HRTFs was verified among three sagittal planes, the lateral angles of which are 0°, 30°, and 60°, as shown in Fig. 5.4.

Figure 5.5 shows the HRTFs of the KEMAR dummy head for the directions for which the vertical angles are identical in the three sagittal planes. Characteristics common among the notches and peaks were observed in the HRTFs, the vertical angles of which are identical, regardless of the lateral angle.

In order to examine the similarity of these HRTFs, correlation coefficients between the amplitude spectra of the HRTFs, the lateral angles of which were different but the vertical angles were same (Table 5.1). For the vertical angles from 0° to 120°, correlations among the HRTFs, the vertical angles of which were

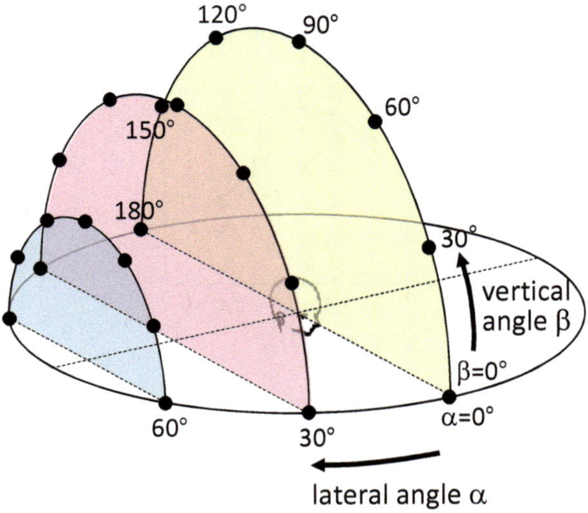

Fig. 5.4 Three sagittal planes and seven vertical angles in each sagittal plane for which the HRTFs were measured

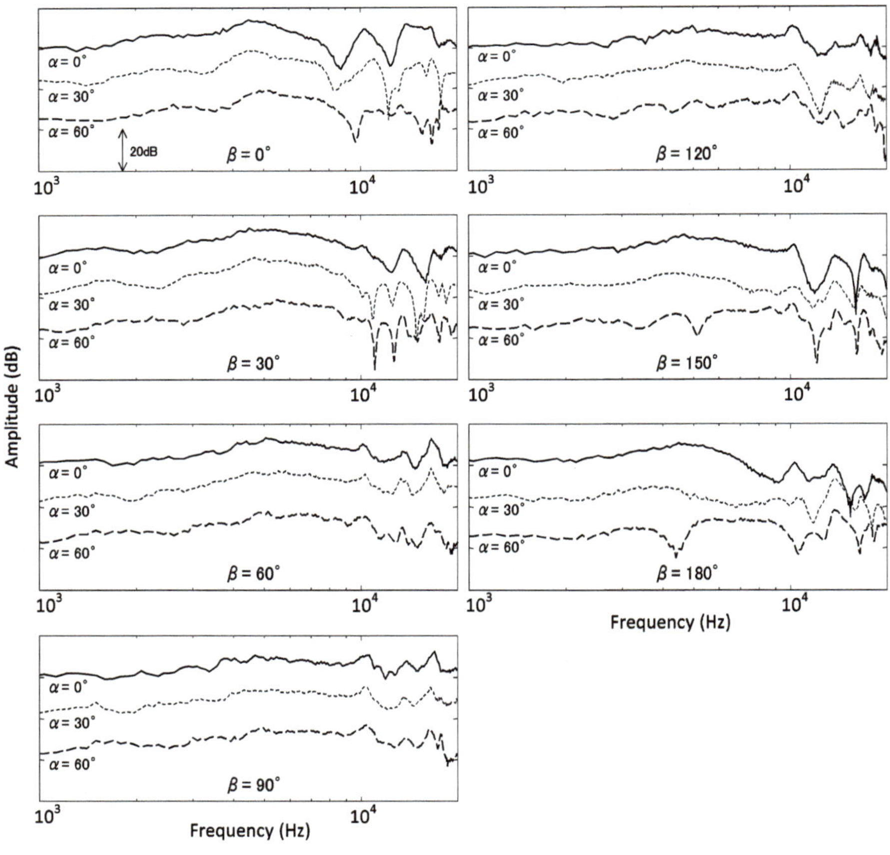

Fig. 5.5 Amplitude spectra of the HRTFs of the KEMAR dummy head. Solid lines, dotted lines, and broken lines indicate the HRTFs for lateral angles of 0° (median plane), 30°, and 60°, respectively

Table 5.1 Correlation coefficients between the amplitude spectra of the HRTFs for which the lateral angles were different but the vertical angles were same

	Combination of lateral angles α (deg.)		
Vertical angle β(°)	0 and 30	0 and 60	30 and 60
0	0.88**	0.53**	0.71**
30	0.92**	0.86**	0.87**
60	0.96**	0.93**	0.94**
90	0.94**	0.87**	0.92**
120	0.91**	0.77**	0.68**
150	0.82**	0.15	−0.19
180	0.54**	−0.11	−0.35

**$p < 0.01$, *$p < 0.05$

same, were high even in the three different sagittal planes. The null hypothesis that there was no correlation was rejected ($p < 0.01$).

On the other hand, for vertical angles of 150° and 180°, significant correlation was observed between the sagittal planes of the lateral angles of 0° and 30°, whereas no correlation was observed between 0° and 60° or between 30° and 60°.

In other words, the amplitude spectra of the HRTFs are considered to be common for the front and upper directions, regardless of lateral angle. For the rear direction, however, the spectral difference was observed among sagittal planes, the lateral angles of which exceed 30°.

5.3 Three-Dimensional Sound Image Control Using the Median Plane HRTFs and Interaural Differences

The results shown in the previous section suggest a localization model in which spectral cues in one sagittal plane (e.g., the median plane) are sufficient for vertical angle perception in any sagittal plane, and sound image localization to an arbitrary three-dimensional direction can be achieved by adding the interaural difference cues to the spectral cues in one sagittal plane is derived. In this section, the validity of the model is examined through sound localization tests.

5.3.1 Three-Dimensional Sound Image Control Using the Measured HRTFs in the Median Plane and Interaural Differences

Figure 5.6 shows the results of the sound localization tests in the upper hemisphere using measured HRTFs in the median plane and the interaural differences (Morimoto et al. 2003a). ITDs and ILDs were obtained from the HRTFs of the four directions in the horizontal plane, the lateral angles of which were 0°–90° (in 30° steps).

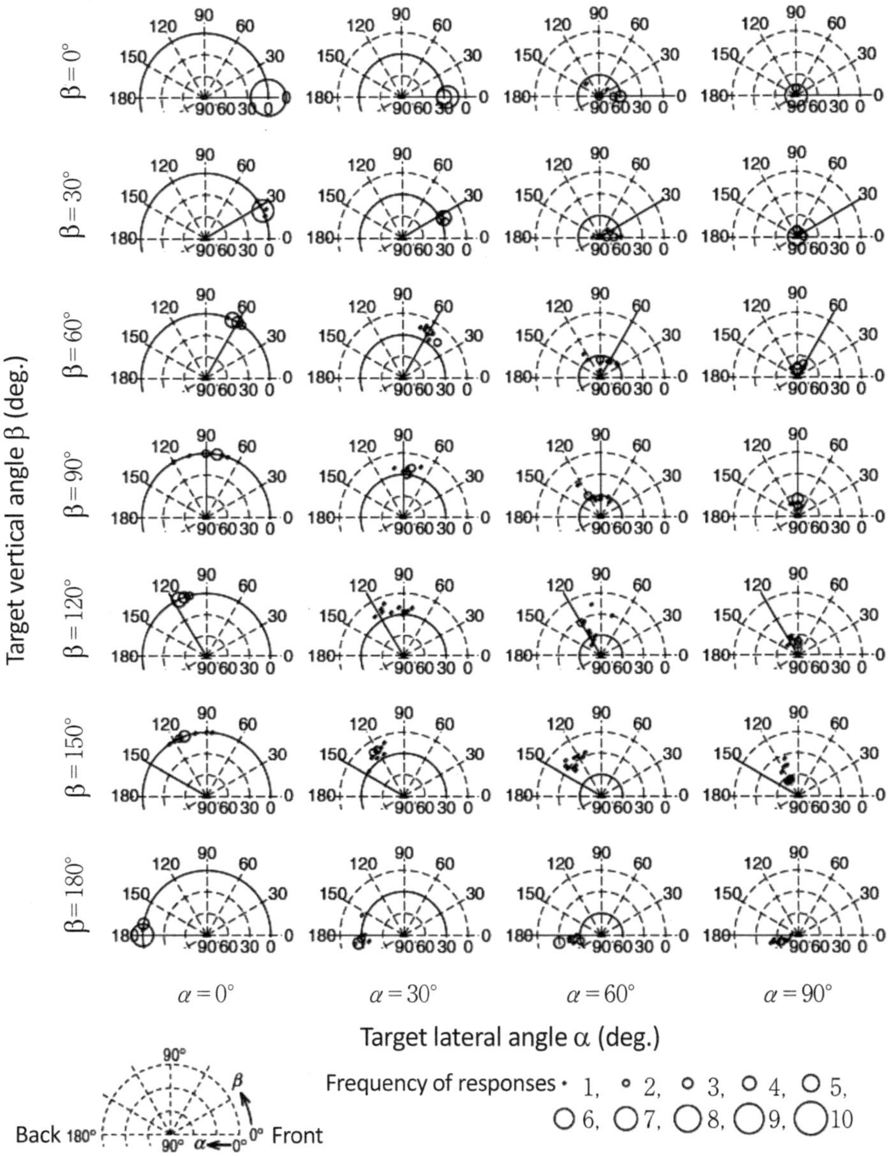

Fig. 5.6 Responses to the stimuli that reproduced the HRTF in the median plane and the interaural differences. The circular arcs denote the lateral angle α, and the straight lines denote the vertical angle β. Bold lines show the target angles α and β. (Morimoto et al. 2003a)

The circular arcs and the straight lines denote the lateral angle and the vertical angle, respectively. The target lateral angles and vertical angles are indicated by bold lines. The responses of the lateral angles distributed at almost the target azimuth angles, while responses were slightly spread at lateral directions. For vertical angles,

Table 5.2 Mean localization errors in lateral angle α and vertical angle β for three sagittal planes using the HRTFs of the median plane and the interaural differences (deg.). (Morimoto et al. 2003a)

	Lateral angle α of the sagittal plane			
	0°	30°	60°	90°
Errors in α	1	7	16	23
Errors in β	15	13	21	–

Table 5.3 Mean localization errors in the lateral angle and the vertical angle for three sagittal planes using the HRTFs of the median plane and the ITD (deg.). (Morimoto et al. 2003a)

	Lateral angle α of the sagittal plane			
	0°	30°	60°	90°
Errors in α	1	13	22	29
Errors in β	13	14	16	–

Table 5.4 Mean localization errors in the lateral angle and the vertical angle for three sagittal planes using the HRTFs of the median plane and the ILD (deg.). (Morimoto et al. 2003a)

	Lateral angle α of the sagittal plane			
	0°	30°	60°	90°
Errors in α	1	15	33	67
Errors in β	15	15	21	–

responses were distributed at the target directions for the front and rear, while responses spread slightly for above directions.

These tendencies are same as those for median plane localization using actual sound sources (see Sect. A.1.2).

Table 5.2 shows the mean localization errors in the lateral angle and vertical angle. For $\alpha = 90°$, the errors were not obtained because β cannot be defined. The mean localization errors in the lateral angle were approximately the same as that for the subject's own HRTF of the target direction (Morimoto and Ando 1980). The mean localization error increases as the target lateral direction increases. This tendency is the same as the increase in the JND of the lateral angle of a sound source with regard to the direction of a sound image (see A1.3). The mean localization errors in the vertical angle were also approximately the same as that for the subject's own HRTF of the target direction.

These results suggest that sound image localization at an arbitrary direction can be achieved using median plane HRTFs combined with interaural differences.

Furthermore, in order to examine the effects of the ITD and ILD on the perception of the lateral angle of a sound image, either only the ITD or the ILD was provided as the interaural difference.

The mean localization errors when only the ITD was provided (Table 5.3) were approximately the same as when both the ITD and ILD were provided (Table 5.2). However, when only the ILD was provided (Table 5.4), the mean localization errors increased compared with the case in which both the ITD and ILD were provided.

It has been reported that the ITD is dominant compared with the ILD for a broadband sound source, which contains low-frequency components, in the perception of lateral angle (Wightman and Kistler 1992). Based on this finding, the information that "ITD is zero" is assumed to act dominantly compared with the ILD when only the ILD was provided. Then, a sound image was shifted to the median plane.

5.3.2 Three-Dimensional Sound Image Control Using the Parametric HRTFs in the Median Plane and Interaural Differences

The above discussion suggests that a sound image can also be controlled to an arbitrary three-dimensional direction using median plane parametric HRTFs, which are recomposed with prominent notches and peaks, combined with interaural differences. The advantage of this method is to focus the problem of individual differences of HRTFs on individual differences in N1 and N2 in the median plane and on the ITDs in the horizontal plane.

Figure 5.7 shows the results of the sound localization tests obtained by this method. The target directions were same as in Fig. 5.4 (22 directions). For the sagittal plane with lateral angle $\alpha = 30°$, the response distribution was approximately the same as for the median plane, while the responses were spread slightly for the sagittal plane with lateral angle $\alpha = 60°$.

The angles between the target directions and the responded directions were calculated as follows, as shown in Fig. 5.8:

$$\theta = \cos^{-1}\frac{\vec{T}\cdot\vec{R}}{\left|\vec{T}\right|\left|\vec{R}\right|} \tag{5.3}$$

where $\vec{T}$ and $\vec{R}$ indicate the vectors of the target direction and the responded direction, respectively.

Table 5.5 shows the mean value of angle θ for two subjects. In the median plane ($\alpha = 0°$) and at the right lateral direction ($\alpha = 90°$), the errors are slightly larger compared to those of measured HRTFs, while the errors are approximately the same in the sagittal plane at $\alpha = 30°$ and $60°$.

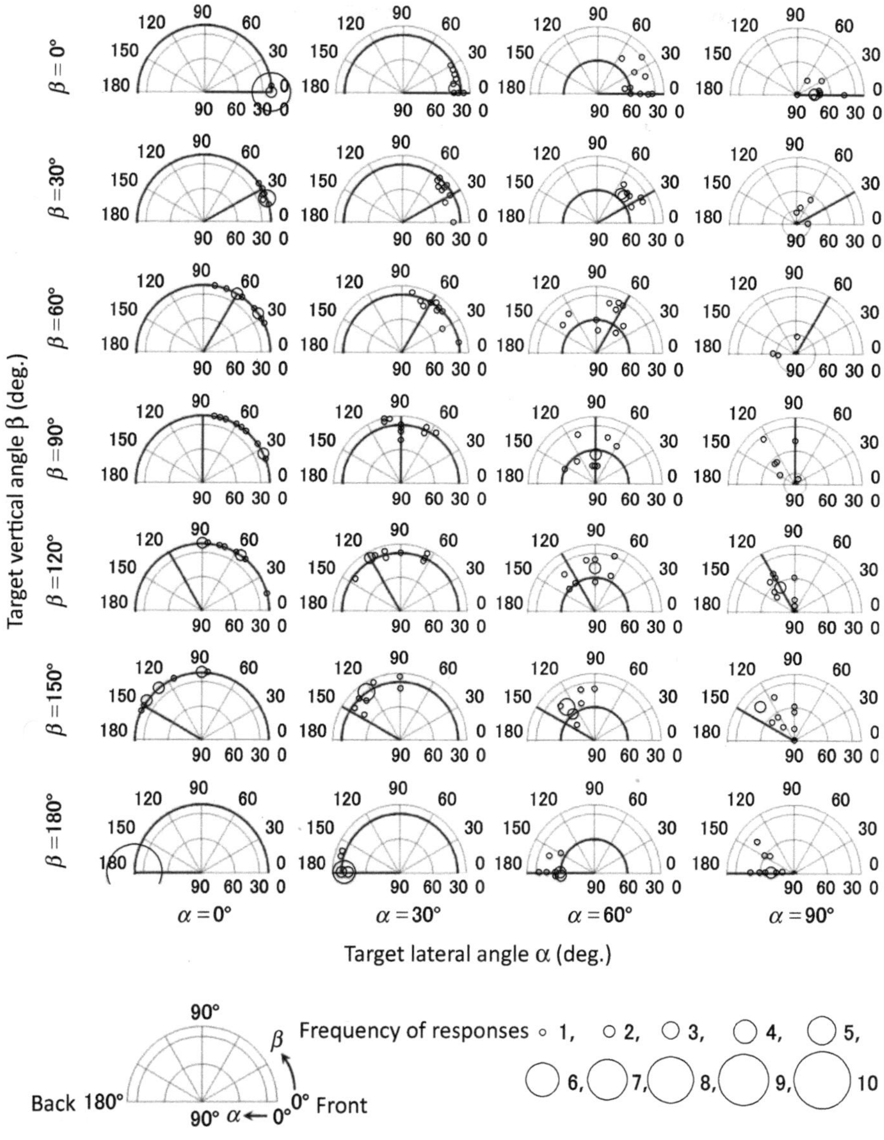

Fig. 5.7 Responses to the stimuli that reproduced the parametric HRTF (N1N2P1) in the median plane and the ITD. The circular arcs denote the lateral angle α, and the straight lines denote the vertical angle β. Bold lines show the target angles α and β

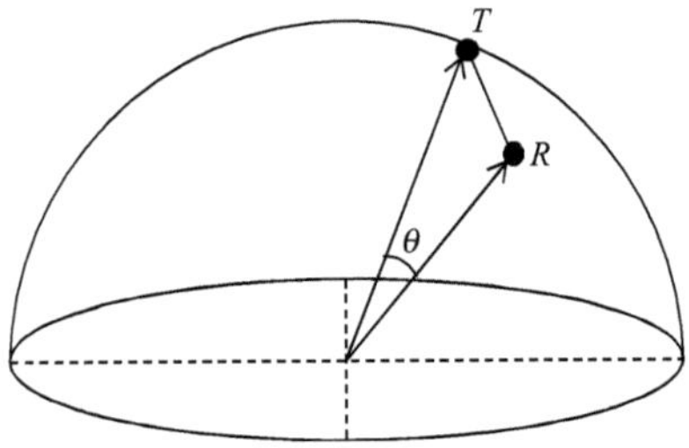

Fig. 5.8 Angle θ between a target direction T and a responded direction R

Table 5.5 Sound image localization error for the parametric HRTF (N1, N2, and P1) in the median plane combined with the ITD for the measured HRTF of the target direction. Errors are indicated by angle θ (deg.) between the target direction and the responded direction (average value of two subjects)

	Lateral angle α (deg.)				
	0	30	60	90	Average
Parametric median plane HRTFs + ITD	21	22	20	14	19
Measured HRTFs	14	21	20	7	16

5.3.3 *Three-Dimensional Sound Image Control Using the Best-Matching HRTF and the ITD*

Three-dimensional sound image control using the best-matching HRTF in the median plane and the ITD were examined. The results of the tests are shown in Figs. 5.9 and 5.10. In these figures, (a) indicates the results for the subject's own HRTF for the target direction, (b) indicates the results for the subject's own HRTF in the median plane combined with the ITD of the horizontal plane, and (c) indicates the results for the best-matching HRTF in the median plane combined with the ITD of the horizontal plane.

For the sagittal plane of $\alpha = 0°$ (median plane) and at $\alpha = 90°$ (just side), the localization accuracy for the best-matching HRTF in the median plane combined with the ITD was approximately the same as that for the subject's own HRTF.

In sagittal planes of $\alpha = 30°$ and $60°$, the sound localization accuracy of the best-matching HRTFs (c) was generally good, but was worse than the subject's own HRTFs for several target vertical angles.

The mean localization errors in the vertical angle and the lateral angle are shown in Tables 5.6 and 5.7, respectively. The mean localization errors in both the vertical angle and the lateral angle for the best-matching HRTF combined with the ITD were approximately the same as those for the subject's own HRTFs of the target direction for most of the target directions. However, in the sagittal planes of $\alpha = 30°$ and $60°$, the errors in vertical angle at $\beta = 90°$ and the errors in lateral angle at $\beta = 180°$ were approximately twice the errors for the subject's own HRTFs, respectively.

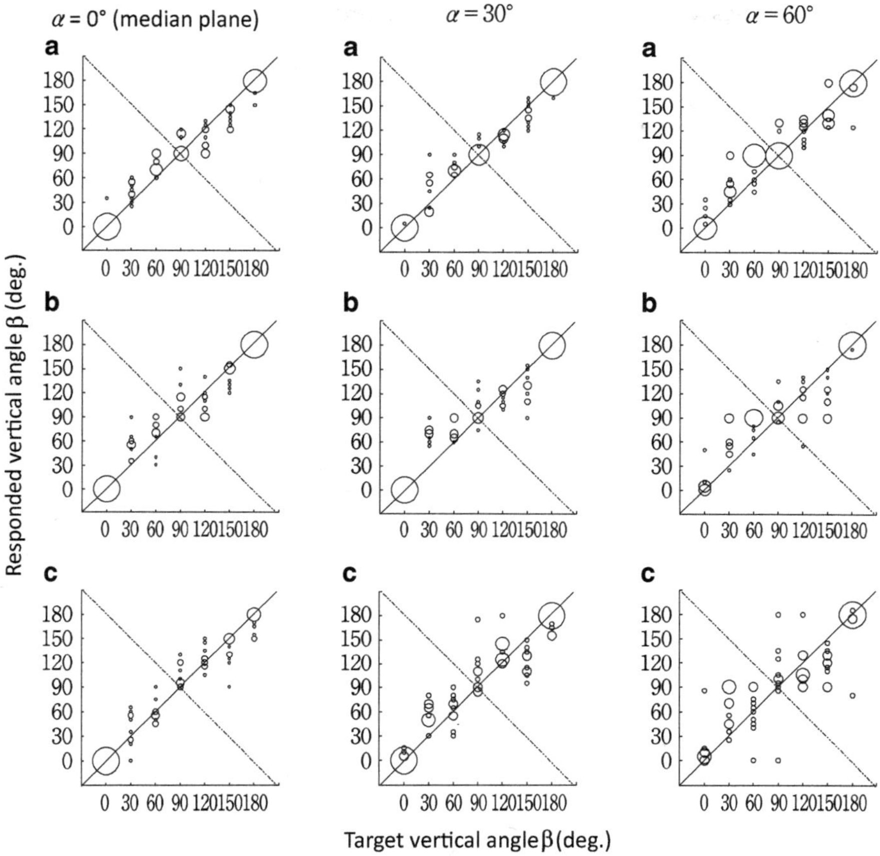

Fig. 5.9 Responses of vertical angle β. (**a**) Subject's own HRTF for the target direction, (**b**) subject's own HRTF in the median plane combined with the ITD of the horizontal plane, and (**c**) best-matching HRTF in the median plane combined with the ITD of the horizontal plane

5.4 Summing Localization Between Sagittal Planes

The summing localization when the identical signal is provided from two loudspeakers located in the horizontal plane has been described in Chap. 2.

Here, as shown in Fig. 5.11, we discuss the sound image when the identical signal is provided from two points S1 and S2, which have same vertical angles in different sagittal planes.

Sound image localization tests were performed in an anechoic chamber using loudspeakers installed in seven directions (30° steps) with a vertical angle β of from 0° to 180° in each sagittal plane at lateral angles of α = 0° (median plane), 30°, and 60° (Morimoto et al. 2003b). A broad-band noise was presented simultaneously from two loudspeakers with equal vertical angles of the sagittal plane of α = 0° and

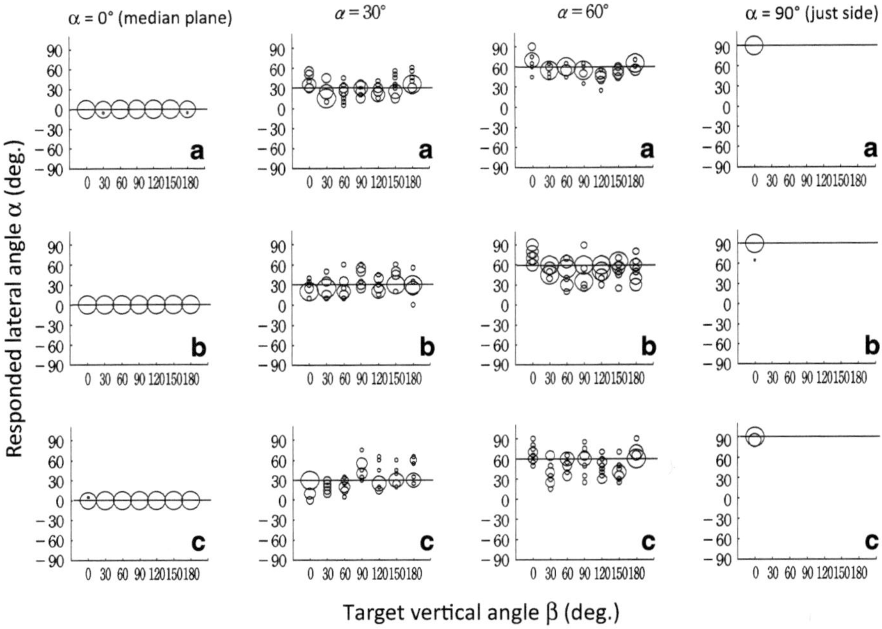

Fig. 5.10 Responses of lateral angle α. (**a**) subject's own HRTF for the target direction, (**b**) subject's own HRTF in the median plane combined with the ITD of the horizontal plane, and (**c**) best-matching HRTF in the median plane combined with the ITD of the horizontal plane

Table 5.6 Mean localization error in vertical angle β (deg.)

Target lateral angle α (deg.)	HRTF	Target vertical angle β (deg.)							
		0	30	60	90	120	150	180	Average
0	(a)	1.9	10.6	15.1	18.4	18.1	11.1	2.2	11.1
	(b)	2.0	19.4	17.7	24.3	16.8	15.4	1.4	13.9
	(c)	1.5	15.8	16.8	16.4	16.4	18.3	4.4	12.8
30	(a)	4.1	20.4	19.6	11.4	13.0	16.3	4.1	12.7
	(b)	2.1	23.3	16.9	19.4	15.2	14.9	0.8	13.2
	(c)	3.8	23.2	18.2	24.3	17.7	16.2	3.2	15.2
60	(a)	15.8	29.0	24.8	9.8	29.2	41.6	9.5	22.8
	(b)	17.2	25.6	20.8	14.9	20.1	30.8	5.5	19.3
	(c)	13.3	28.3	19.2	23.9	21.6	31.9	11.9	21.4

(a) Subject's own HRTF for the target direction, (b) subject's own HRTF in the median plane combined with the ITD of the horizontal plane, and (c) best-matching HRTF in the median plane combined with the ITD of the horizontal plane

60°, as shown in Fig. 5.12. For comparison, the broad-band noise was presented from a single loudspeaker in the sagittal plane of α = 30°.

Figures 5.13(a) shows the responses with regard to the lateral angle and vertical angle when a broad-band noise was presented from a single loudspeaker in the sagittal plane of α = 30°, respectively. The responses with regard to the lateral angle

Table 5.7 Mean localization error in lateral angle α (deg.)

Target lateral angle α (deg.)	HRTF	Target vertical angle β (deg.)							
		0	30	60	90	120	150	180	Average
0	(a)	4.5	1.6	0.8	0.9	0.6	0.3	2.5	1.6
	(b)	4.9	2.6	0.9	0.8	0.3	0.3	8.2	2.6
	(c)	0.6	0.3	0.3	0.0	1.2	0.6	2.5	0.8
30	(a)	7.6	12.3	10.2	11.9	9.9	10.4	10.3	10.4
	(b)	10.6	15.2	18.4	15.0	15.2	13.1	11.2	14.1
	(c)	11.4	12.6	12.7	16.2	16.3	13.1	20.9	14.7
60	(a)	10.0	13.5	7.2	8.8	14.9	13.5	9.2	11.0
	(b)	13.8	17.3	17.9	19.3	16.8	11.2	9.5	15.1
	(c)	11.3	18.1	13.8	11.9	13.8	18.3	18.3	15.1
90	(a)	2.1							
	(b)	3.3							
	(c)	2.3							

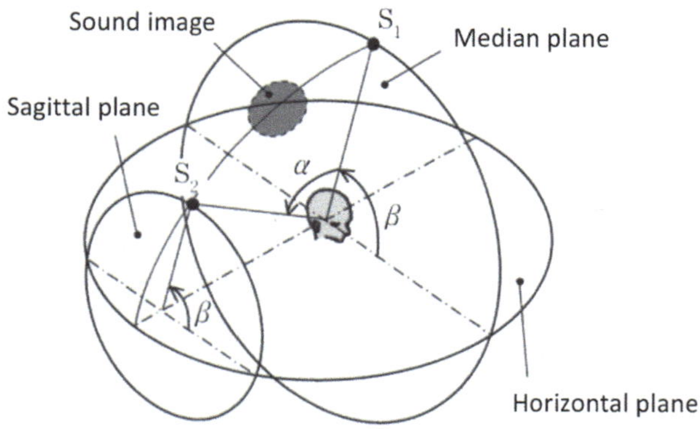

Fig. 5.11 Two points S1 and S2 that have same vertical angles in different sagittal planes

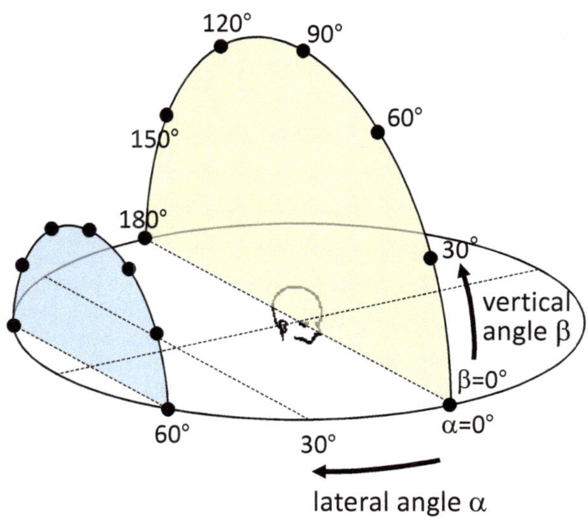

Fig. 5.12 Loudspeaker arrangements for tests for summing localization

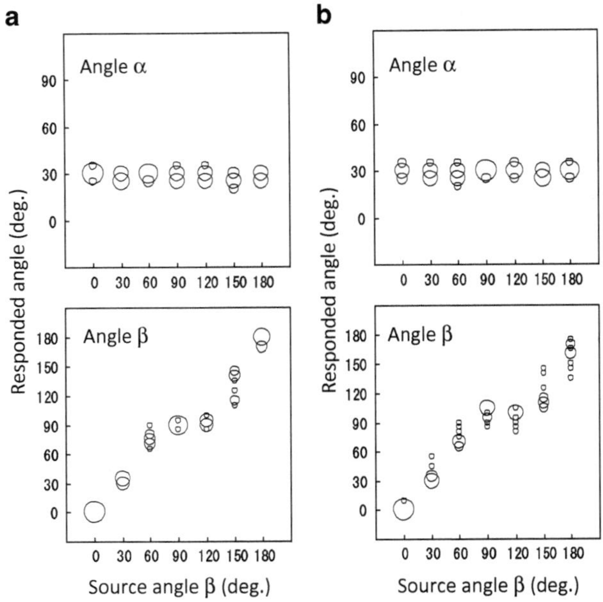

Fig. 5.13 Responses to broad-band noise presented simultaneously from a single loudspeaker in the sagittal plane of α = 30° (**a**) and from two loudspeakers with equal vertical angles in the sagittal planes of α = 0° and 60° (**b**) (Morimoto et al. 2003b)

were distributed around 30°, and most of the responses with regard to the vertical angle were distributed over a diagonal line.

Figures 5.13(b) shows the responses with regard to lateral angle and vertical angle when the broad-band noise was presented simultaneously from two loudspeakers with equal vertical angles of the sagittal plane of α = 0° and 60°, respectively. The responses with regard to lateral angle were all distributed around 30°. The lateral angle was perceived between the two loudspeakers, as in the case of the summing localization in the horizontal plane. The responses with regard to vertical angle were distributed over a diagonal line.

These results suggest that a sound image is perceived between the two sound sources when an identical signal is provided from the two loudspeakers, which have the same vertical angles in different sagittal planes.

References

Blauert J (1969/70) Sound localization in the median plane. Acust 22: 205–213

Iida K, Ishii Y (2011) Individualization of the HRTFs in the basis of the spectral cues for sound localization. In: Suzuki Y, Brungard D, Iwaya Y, Iida K, Cabrera D, Kato H (eds) Principles and applications of spatial hearing. World Scientific, Singapore, pp 159–178

Morimoto M, Ando Y (1980) On the simulation of sound localization. J Acoust Soc Jpn (E) 1:167–174

Morimoto M, Aokata H (1984) Localization cues of sound sources in the upper hemisphere. J Acoust Soc Jpn (E) 5:165–173

Morimoto M, Iida K, Itoh M (2003a) Upper hemisphere sound localization using HRTFs in the median plane and interaural differences. Acoust Sci Tech 24:267–275

Morimoto M, Itoh M, Iida K (2003b) Localization of sound image produced by two sound sources in sagittal planes. Proc. 8th Western Pacific Acoustics Conference (WESPAC VIII), TE13 (4 pages), Melbourne, Australia, 7–9 April

Nishino T, Kajita S, Takeda K, Itakura F (2001) Interpolation of head related transfer functions of azimuth and elevations. J Acoust Soc Jpn 57:685–692. in Japanese

Wightman FL, Kistler DJ (1992) The dominant role of low–frequency ITD in sound localization. J Acoust Soc Am 91:1648–1661

Chapter 6
Directional Band and Spectral Cue

Abstract Blauert (ACUSTICA 22: 205–213, 1969/70) performed sound localization tests, in which 1/3 octave band noise was presented randomly from the front, above, and rear in the median plane and reported that there was a band that was perceived in a specific direction, regardless of the direction of the sound source. This band is referred to as the directional band.

6.1 Directional Band

Figure 6.1 shows the relative frequency of the subjects, the number of responses in one of the front, upper, or rear direction is considered to be greater than the number of responses in the other two directions at a significance level of 5%.

The open square in the figure indicates a directional band, which can be regarded as having more subjects, who perceived a sound image to be in that direction at a significance level of 10%. The hatched square indicates a quasi-directional band.

These results show that a 1/3 octave band with center frequencies of 315–500 Hz and 3.15–5 kHz is perceived as being forward, with center frequencies of 800–1.6 kHz and 10–12.5 kHz as being backward, and with a center frequency of 8 kHz as being upward.

Furthermore, Blauert analyzed the HRTFs and reported that the energy in the directional band was large as compared to other directions. He referred to this band as the boosted band.

6.2 Individual Differences in Directional Bands

Itoh et al. (2007) reported that there exist individual differences in directional bands. The directional bands obtained for each of the seven subjects are shown in Table 6.1. Here, "All" indicates the directional band obtained from the responses of all of the subjects. The directional band obtained by Blauert (1969/70) is also shown.

K. Iida, *Head-Related Transfer Function and Acoustic Virtual Reality*,
https://doi.org/10.1007/978-981-13-9745-5_6

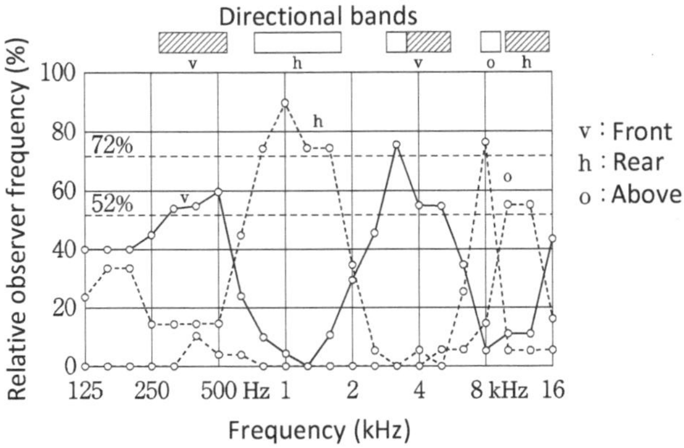

Fig. 6.1 Directional bands. (Blauert 1969/70)

It is common that the directional band changes as the center frequency becomes higher in any of the subjects: rear → front → above → rear. However, there is a difference in the frequency. That is, there are individual differences in the directional band. On the other hand, no significant difference is observed between the 1/3 octave band noise and the 1/6 octave band noise.

6.3 Band Widths of Directional Bands

The band widths in which directional bands occur have also been examined. Table 6.2 shows the directional bands for the 1/6 octave band noise for four subjects.

For the three types of center frequencies, in which directional bands occurred (rear: 1250 Hz, front and rear: 4000 Hz, above and rear: 7100 Hz), sound image localization tests were conducted using stimuli (1/6, 1/12, and 1/24 octave bands and pure tone), for which the band width was narrowed.

Table 6.3 shows the results. With some exceptions, by narrowing the bandwidth, even for pure tones, directional bands similar to 1/6 octave bands occurred.

Furthermore, sound image localization tests were performed using stimuli, in which successive directional bands of 1/6 octave bands occur in the same direction, were connected. The results are shown in Table 6.4.

Directional bands occurred in the direction same as the 1/6 octave bands for four of five types of stimuli for Subject SKG. However, for the stimuli having a center frequency from 11,224 Hz to 16,000 Hz, a directional band occurred in the rear, whereas that for the 1/6 octave band occurred in the front. For Subject NMR, the same directional band as the 1/6 octave band occurred for all three types of stimuli.

These results infer that, even for the case in which the band width of the stimulus is widened by connecting consecutive directional bands, which are perceived in the same direction, the directional bands are preserved.

Table 6.1 Directional bands for seven subjects. (Itoh et al. 2007)

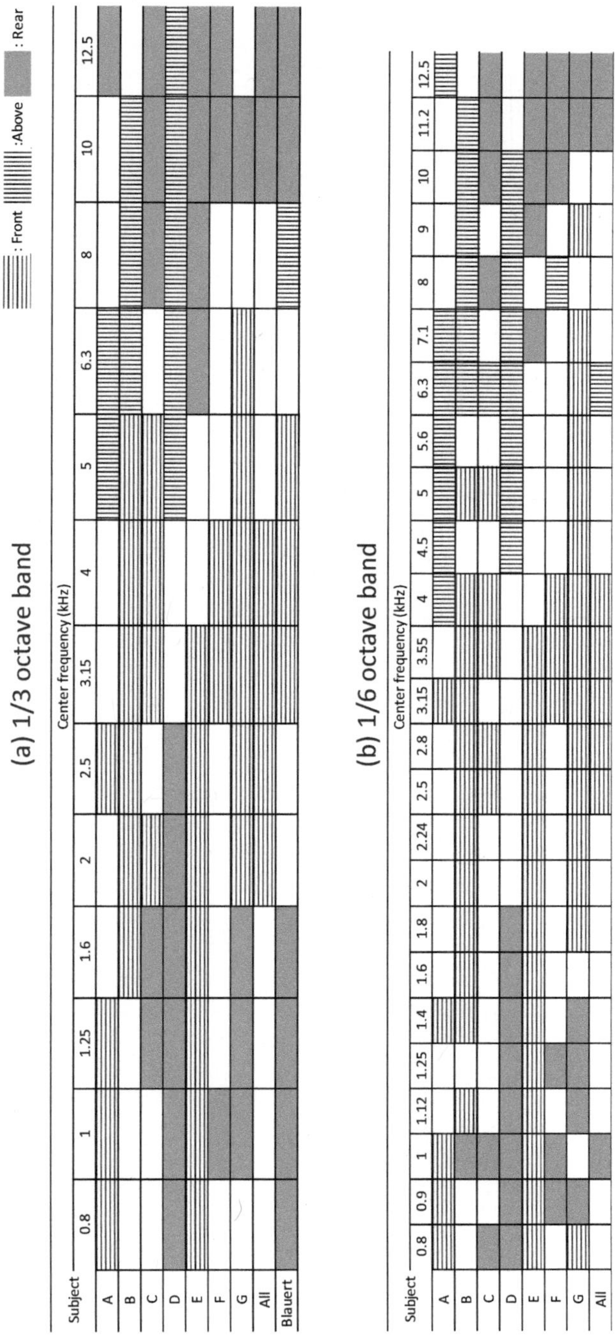

Table 6.2 Directional bands for the 1/6 octave band

[horizontal lines]: Front [vertical lines]: Above [grey]: Rear

Subject	Center frequency (Hz)																	
	250	280	315	355	400	450	500	560	630	710	800	900	1000	1120	1250	1400	1600	1800	2000
SKG	Front	Front	Front	Front	Front	Front	Front	Front	Front	Front	Front	Front	Front	Rear	Rear	Rear	Rear	Rear	Rear
KMG	Rear	Rear	Rear	Rear	Rear	Rear	Rear	Rear	Rear	Rear	Rear	Rear	Rear	Rear	Rear	Rear	Rear	Rear	Rear
MYM															Rear	Rear			
NMR	Rear	Front	Rear			Rear	Rear	Rear	Rear	Rear	Rear	Rear	Rear	Rear	Rear	Rear	Rear	Rear	Rear

Subject	Center frequency (Hz)																	
	2250	2500	2800	3150	3550	4000	4500	5000	5600	6300	7100	8000	9000	10000	11200	12500	14000	16000
SKG	Rear	Rear	Rear	Rear		Front	Front			Above	Above	Above	Above	Rear	Front	Front	Front	Front
KMG	Rear	Rear	Rear	Rear	Rear	Rear	Rear	Rear	Rear	Rear	Rear		Rear	Rear	Rear	Rear	Rear	Rear
MYM											Above							
NMR	Rear	Rear	Front	Front	Front	Front	Front	Front	Front		Above	Above	Above	Above		Front		

Table 6.3 Directional bands for the narrow band

(horizontal lines): Front (vertical lines): Above (grey): Rear

(a) 1,250 Hz

Subject	Band width			
	1/6 octave	1/12 octave	1/24 octave	pure tone
SKG	Rear	Front	Rear	Rear
KMG	Rear	Rear	Rear	Rear
MYM	Rear	Rear	Rear	Rear
NMR				

(b) 4,000 Hz

Subject	Band width			
	1/6 octave	1/12 octave	1/24 octave	pure tone
SKG	Front	Front	Front	Front
KMG	Rear	Rear	Rear	Rear
MYM		Front		
NMR				Above

(c) 7,100 Hz

Subject	Band width			
	1/6 octave	1/12 octave	1/24 octave	pure tone
SKG	Above	Above	Above	Above
KMG	Above	Above	Above	Above
MYM	Above	Above	Above	Above
NMR				

Table 6.4 Directional bands for connecting consecutive bands

(horizontal lines): Front (vertical lines): Above (grey): Rear

(a) Subject SKG

	Frequency (Hz)				
	250 - 1000	1120 - 3150	4000 - 4500	6300 - 9000	11200 - 16000
1/6 octave band	Front	Rear	Front	Above	Front
Connected Band	Front	Rear	Front	Above	Rear

(b) Subject NMR

	Frequency (Hz)		
	450 - 2500	2800 - 5600	7100 - 10000
1/6 octave band	Rear	Front	Above
Connected Band	Rear	Front	Above

6.4 Relationship Between Directional Band and Spectral Cue

As described in Sect. 6.3, even for pure tones or the directional bands of continuous 1/6 octave bands, the narrow-band signals are localized at a specific direction.

Then, what kind of sound image will be perceived if the energy of the directional band of the broad-band signal is boosted? As described above, when 1/3 octave band noise is presented from the median plane, most of the subjects localize a sound image above, regardless of the vertical direction of the sound source.

One sound image was localized at the direction of the loudspeaker when broad-band noise, in which the sound pressure in the 1/3 octave band of 8 kHz is boosted (up to +18 dB), was presented from a loudspeaker placed at the front or rear direction. Beyond a boost of +18 dB, only the boosted band spatially separates and is localized above, and the other band is localized at the direction of the loudspeaker.

This ear input signal has information on the sound source direction (front or back) as a notch and has information on the above direction as a boosted band. Therefore, this experimental result suggests that the notch functions more strongly than the boosted band as a spectral cue of vertical angle perception.

Middlebrooks (1992) proposed that the auditory system has the knowledge of the directional information filter by the pinna and the sound image occurs in the direction of the filter that the ear input signal best fits. Based on this proposal and the above experimental results, it is reasonable to consider that, "For vertical angle perception, the auditory system collates the ear input signal with the knowledge of the spectrum of HRTF. The auditory system uses the notch frequency as a stronger cue, and when the notch frequency is not available (such as narrow band signals), the auditory system uses the boosted band".

References

Blauert J (1969/70) Sound localization in the median plane. ACUSTICA 22:205–213

Itoh M, Iida K, Morimoto M (2007) Individual differences in directional bands. Appl Acoust 68:909–915

Middlebrooks JC (1992) Narrow–band sound localization related to external ear acoustics. J Acoust Soc Am 92:2607–2624

Chapter 7
Distance Perception and HRTF

Abstract Reproduction and control of sound image distance are also important elements in three-dimensional acoustics. However, quantitative reproduction and control of sound image distance have not yet been realized. In particular, sound images in the front direction are often localized inside of the head or extremely close to the forehead. In this chapter, the cues for the perception of sound image distance are described. In particular, the relationship between the sound image distance and the incident direction is discussed in detail.

7.1 Sound Source Distance and Sound Image Distance

First, the experimental results on the relationship between the sound source distance and the sound image distance are introduced.

Figure 7.1 shows the relationship between the speaker distance up to 10 m in the front direction and the sound image distance (Von Bekesy 1949). Note that, in this figure, the horizontal axis is the sound image distance, and the vertical axis is the sound source distance. The sound image distance coincides with the sound source distance up to approximately 3 m, but the sound image distance does not increase greatly, even if the sound source distance increases.

In other words, the sound image does not occur over a wide range of distance, and there is a limitation for the auditory space in which the sound image distance is perceived.

Why does such a phenomenon occur? As described above, in direction perception, an HRTF, the characteristics of which change significantly depending on the sound source direction, is an important cue. However, the HRTF depends on the distance only in the near sound field within approximately 1 m of the sound source and changes only slightly at greater distances.

In other words, for a sound source at a distance of 1 m or more, the HRTF does not become a cue for distance perception. The absence of "physical cues derived from the human body" reflecting the difference in sound source distance makes distance perception difficult.

The original version of this chapter was revised. The correction to this chapter is available at https://doi.org/10.1007/978-981-13-9745-5_14

K. Iida, *Head-Related Transfer Function and Acoustic Virtual Reality*,
https://doi.org/10.1007/978-981-13-9745-5_7

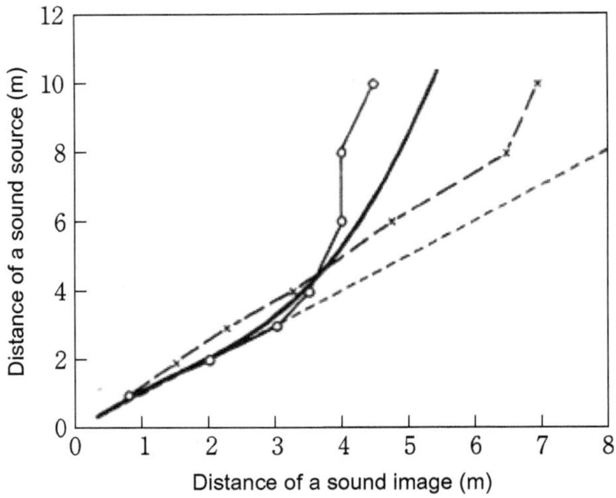

Fig. 7.1 Relationship between the distance from a listener to a speaker and the sound image distance perceived by the listener (Von Bekesy 1949). Open circles and x symbols denote the responses of two subjects. The bold line denotes the average of five subjects. The subjects were blindfolded

7.2 Physical Characteristics that Affects Sound Image Distance

In the process of propagating sound waves in space, there are several physical characteristics that affect the sound source distance. The principal characteristics are described below.

7.2.1 Sound Pressure Level

Keeping the output sound pressure of the sound source constant and changing the distance from the sound source, the sound pressure level at the receiving point changes, and as a result, the loudness also changes.

Figure 7.2 shows subjects' responses to the sound image distance obtained in an experiment in which five loudspeakers were arranged in a line at equal intervals between 3 m and 9 m in front of the subject and two loudspeakers at 3 m and 9 m emitted voices at various sound pressure levels (Gardner 1969). The sound image distance is independent of the actual distance of the sound source and depends on the sound pressure level at the position of a subject. Similar results have been obtained in a number of studies, and it is clear that the sound pressure level at the listener's position affects the sound image distance.

Here, let us consider the mechanism by which the sound pressure level is related to the sound image distance. In order for the sound pressure level to be a cue for distance perception, it is necessary for the listener to have knowledge of the listening sound pressure level or the loudness of the target sound source at a certain distance in advance. Since this condition is not always satisfied, it is not always possible to accurately perceive the distance of the sound source based on the sound pressure level.

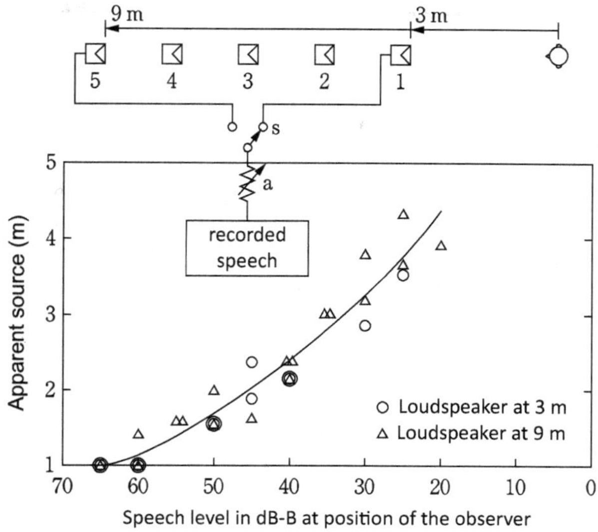

Fig. 7.2 Relationship between the sound pressure level at the listener and the sound image distance for loudspeakers placed at 3 m and 9 m. (Gardner 1969)

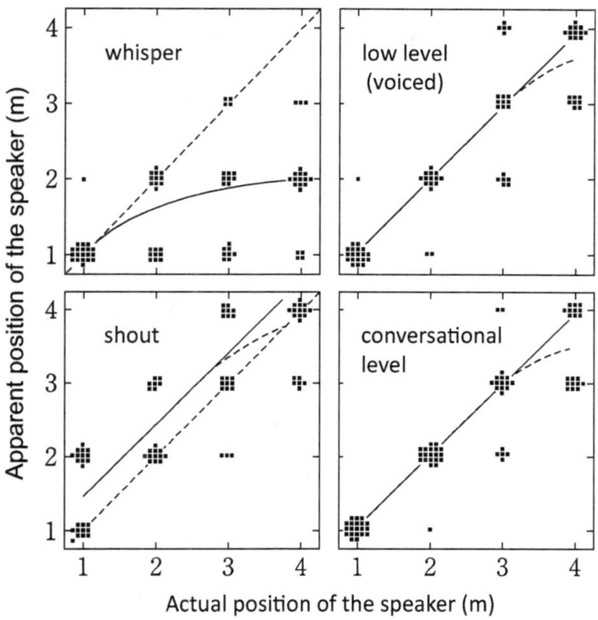

Fig. 7.3 Relationships between the sound source distance and the sound image distance for different types of live voice. (Gardner 1969)

Figure 7.3 shows the relationships between the sound source distance and the sound image distance in an anechoic room for different types of live voice, such as "whisper", "shout", "low level", and "conversation level". Since the subject was blindfolded, there was no visual information on the sound source distance. This figure shows that even for the same sound source distance, the sound image distance for "shout" is farther than that of "conversation", and that of "whisper" is closer than

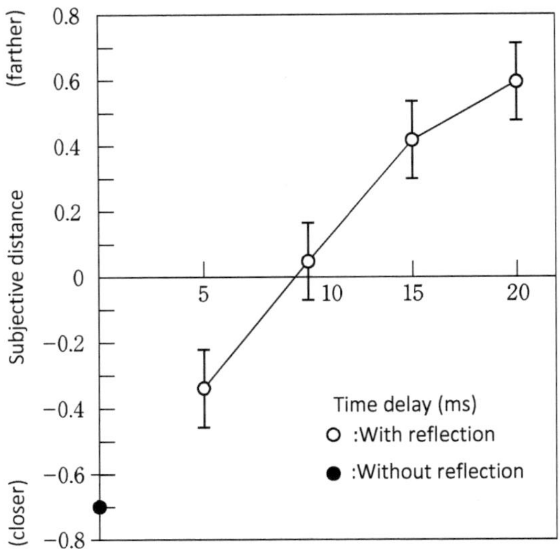

Fig. 7.4 Relationship between the time delay of a single reflection and the sound image distance. (Gotoh et al. 1977)

that of "conversation". This suggests that humans learn the relationship between the sound source distance and the listening sound pressure level or loudness for each type of sound source and use this value for distance perception.

7.2.2 *Time Delay of Reflections*

In a usual sound field, in addition to direct sound, many reflected sounds are incident. It has been reported that sound image distance increases with an increase in the delay time of a reflection, as shown in Fig. 7.4 (Gotoh et al. 1977).

Furthermore, experiments in which the reflections of the actual sound field were simulated while changing the distance between the sound source and the receiving point were performed. The results showed that sound image distance was perceived in the order of the distance between the sound source and the receiving point.

These results suggest that humans use reflections as a cue for distance perception.

7.2.3 *Incident Direction*

A. *Relationship between incident azimuth angle and sound image distance*

The sound image distance is also influenced by the incident direction. Experiments were conducted in which white noise was presented in pairs to a subject from loudspeakers placed at twelve azimuth angles in the horizontal plane (in 30° steps) in an anechoic chamber.

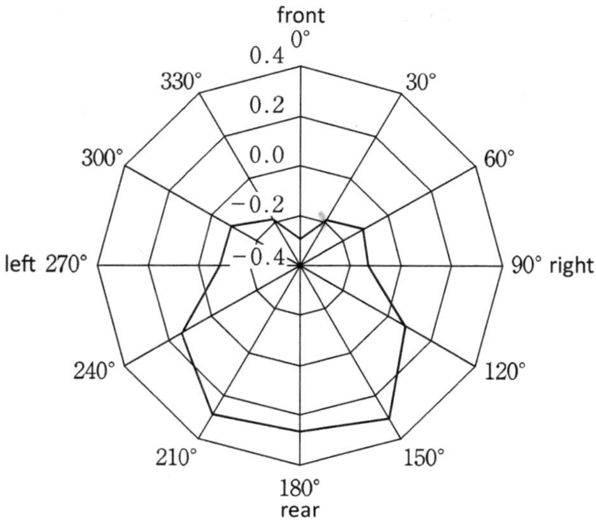

Fig. 7.5 Relationship between the azimuth angle of a sound source and the sound image distance

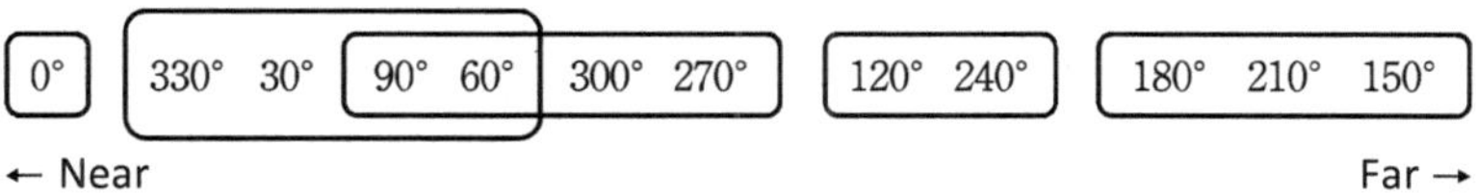

Fig. 7.6 Groups of azimuth angles, in which there exists no significant difference in sound image distance ($p < 0.01$)

Figure 7.5 shows the sound image distance obtained by Ura Variation of Scheffe's paired comparison. The curve is concave at the front and broadens as it moves backward. That is, the sound images at 0° and ± 30° are close, and the sound images at 180° and ± 150° are far.

Figure 7.6 shows groups of azimuth angles, in which there exists no significant difference in sound image distance ($p < 0.01$). The twelve azimuth angles were divided into five groups: front, diagonally front, sideways, diagonally back, and rear. The sound sources located at symmetrical positions are in the same group, and the distance perception of a sound source in the horizontal plane is considered to have left-right symmetry.

Figure 7.7 shows relationship between the azimuth angle and the sound image distance for each subject. There exist some individual differences in the incident azimuth angle dependence of the sound image distance.

Figure 7.8 shows groups of azimuth angles in which there exist no significant difference in sound image distance ($p < 0.01$). The azimuth angle, at which the sound image distance was the nearest, was 0° for seven out of ten subjects. The azimuth angle at which the sound image distance was the farthest was 150°, −150°, or 180° for eight out of ten subjects. This tends to be the same as the average sound image distance for all subjects.

Although the relationship between the incident azimuth angle of the sound source and that of the sound image distance differ somewhat depending on the subject, the

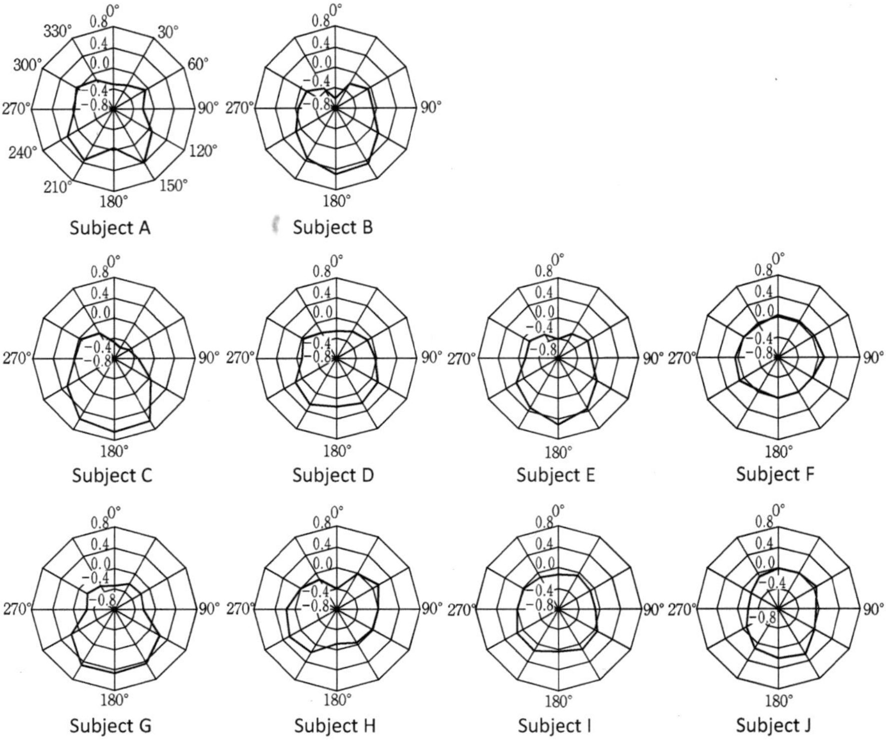

Fig. 7.7 Relationship between the azimuth angle of a sound source and the sound image distance for each subject

tendency to perceive a front sound image as being near and a rear sound image as being far is common. However, subject F did not differ significantly for all pairs.

B. *Relationship between incident vertical angle and sound image distance*

The sound image distance was obtained for seven vertical angles in the upper median plane (in 30° steps) for ten subjects as well as for the azimuth angle. The results are shown in Fig. 7.9. Sound images of 0° and 30° were perceived to be near, and those of 120°, 90°, and 150° were perceived to be far.

Figure 7.10 shows groups of vertical angles, in which there exists no significant difference in sound image distance ($p < 0.01$). The seven vertical angles were divided into three groups. The vertical angle at which the sound image distance was the nearest was 0°, and that at which the sound image distance was the farthest was diagonal back.

Figure 7.11 shows relationship between the vertical angle and the sound image distance for each subject. The vertical angle at which the sound image distance was the nearest was 0° or 30° for nine out of ten subjects. All ten subjects responded that 90°, 120°, or 150° was the farthest.

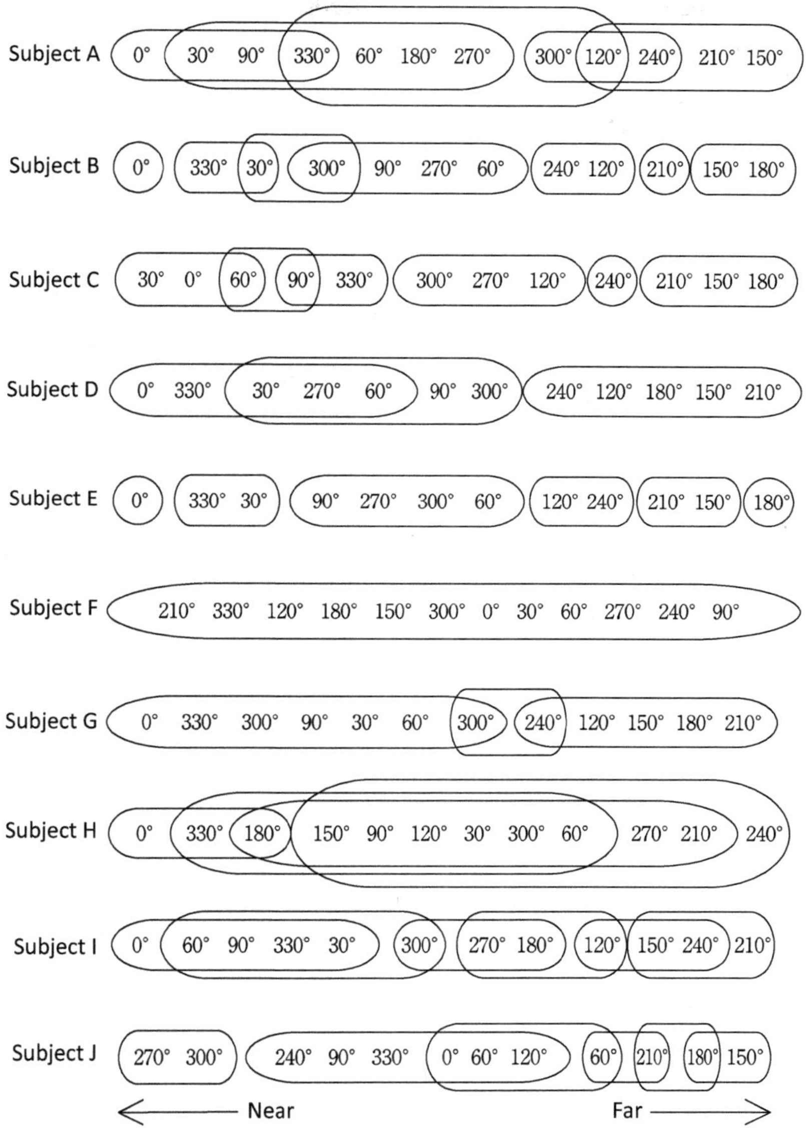

Fig. 7.8 Groups of azimuth angles, in which there exists no significant difference in sound image distance (p < 0.01) for each subject

Figure 7.12 shows groups of vertical angles in which there exists no significant difference in sound image distance ($p < 0.01$). The vertical angle group for which the sound image distance was the nearest included 0° for seven out of ten subjects. The vertical angle group for which the sound image distance was the farthest included 120° for eight out of ten subjects.

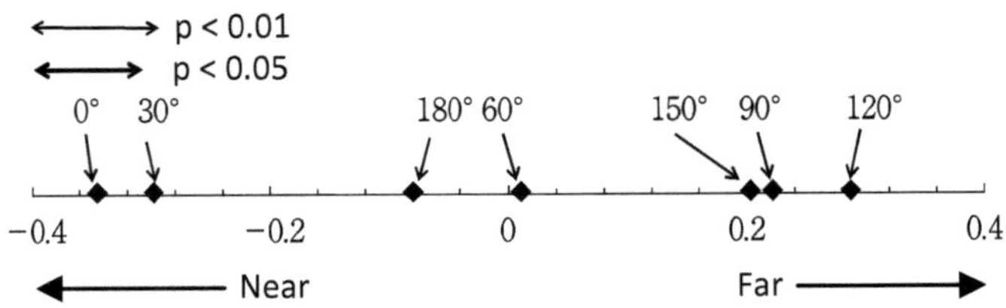

Fig. 7.9 Relationship between the vertical angle of a sound source and the sound image distance

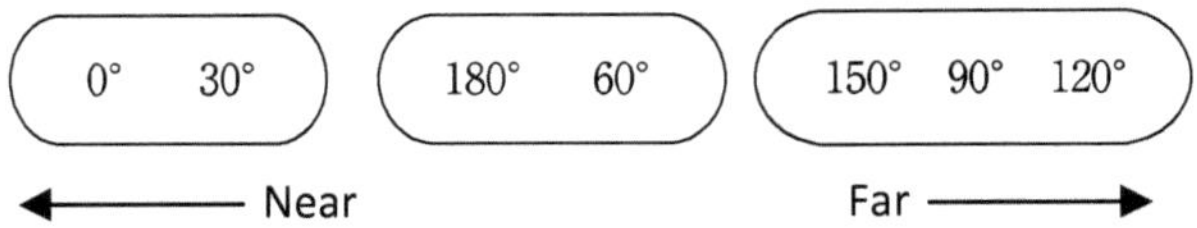

Fig. 7.10 Groups of vertical angles, in which there exists no significant difference in sound image distance ($p < 0.01$)

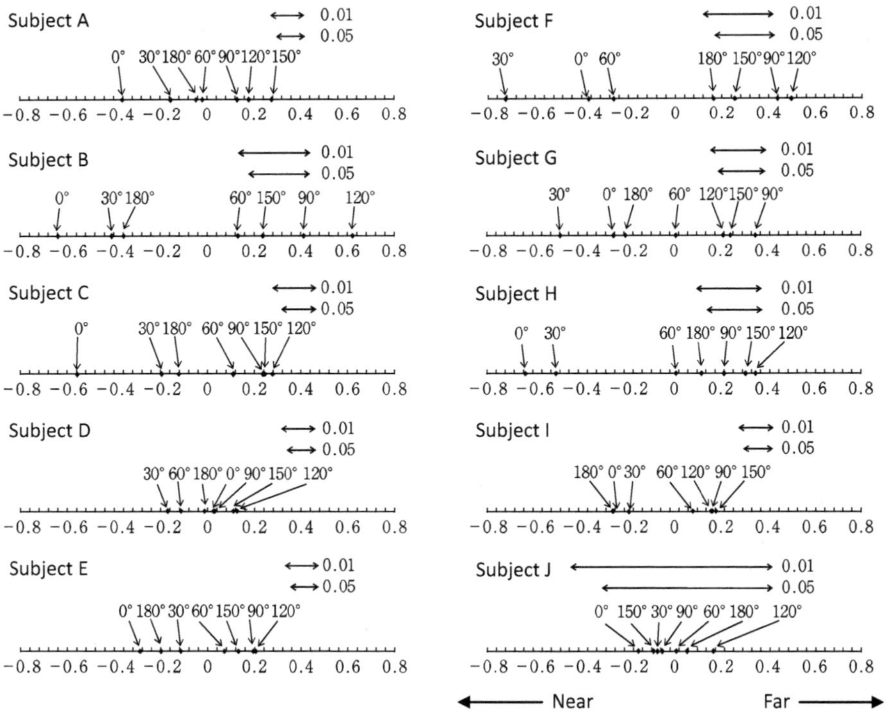

Fig. 7.11 Relationship between the vertical angle of the sound source and the sound image distance for each subject

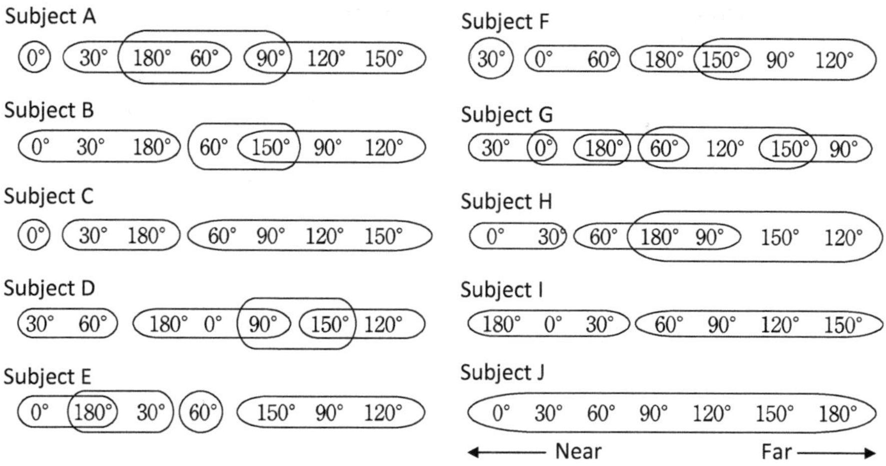

Fig. 7.12 Groups of vertical angles, in which there exists no significant difference in sound image distance ($p < 0.01$) for each subject

These results infer that most of the subjects perceived the front sound images as being near and the diagonally rear sound images as being far. However, subject J did not differ significantly for all pairs.

C. *Relationship between binaural sound pressure level and sound image distance*

One reason for the difference in sound image distance depending on the sound source direction can be considered binaural summation of sound pressure level (BSPL), which is defined as follows (Robinson and Whittle 1960):

$$BSPL = 6\log_2\left(2^{L_l/6} + 2^{L_r/6}\right) \tag{7.1}$$

where L_l and L_r indicate the sound pressure levels of the left ear and the right ear, respectively.

Figure 7.13 shows the average values of all subjects' relative BSPLs in the horizontal plane and the upper median plane, with the BSPL of the front direction set as 0 dB. This figure suggests that the BSPL at the front direction is large in both the horizontal and median planes and there is a negative correlation between the BSPL and the sound image distance.

Figure 7.14 shows the relative BSPL of each subject in the upper median plane. For all subjects, the BSPL at the front direction is large, and, in the eight subjects

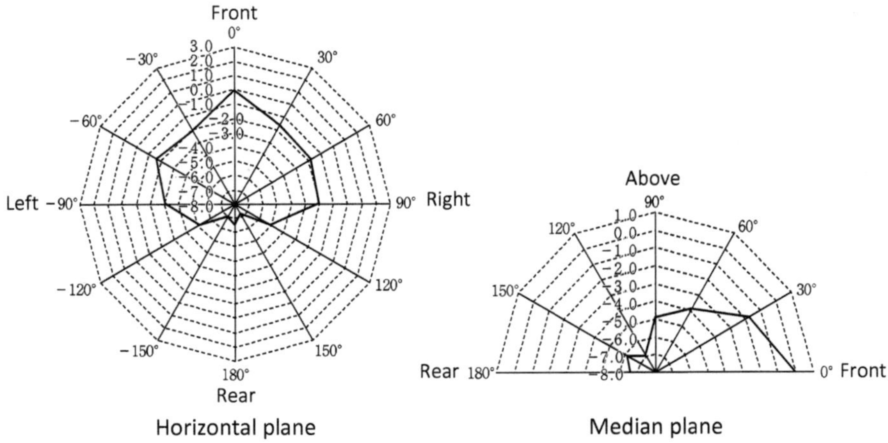

Fig. 7.13 Relative BSPL in the horizontal plane and the upper median plane

excluding subjects G and J, the BSPL at the diagonally back direction is small.

Furthermore, the result of single regression analysis with the relative BSPL as an objective variable and sound image distance as an explanatory variable is shown in Fig. 7.15. Sound image distance tends to be near with an increase in BSPL, with the exception of several subjects.

D. *Summary of the relationship between incident direction and sound image distance*

In summary, the sound image distance differs depending on the incident direction of sound both in the horizontal plane and in the median plane. In the horizontal plane, the sound image distance in the front direction is near, and that in the rear direction (150° to 210°) is far. In the median plane, the sound image distance in the front direction is near, and those in the above and diagonally back directions (90°–150°) are far, as shown in Fig. 7.16.

It has been reported that the "sound image distance of the front direction is perceived nearer than those in other directions" when a three-dimensional acoustic reproduction regeneration system is used. The reason was thought to be that signal processing was not precisely realized. However, the experimental results suggest that the frontal sound image distance is perceived near not only due to a signal processing problem, but also due to human auditory characteristics.

This indicates the possibility that an equidistant sound image is generated by presenting sound so that the BSPL of each direction is the same (Fig. 7.17).

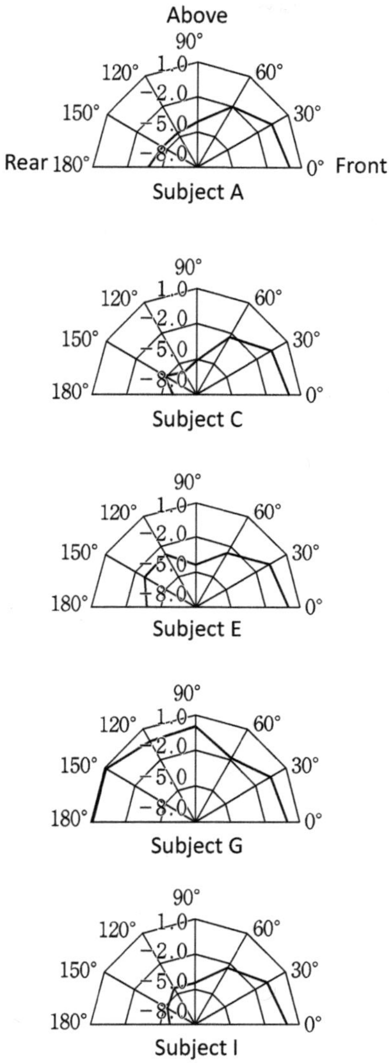

Fig. 7.14 Relative BSPL in the upper median plane for each of ten subjects

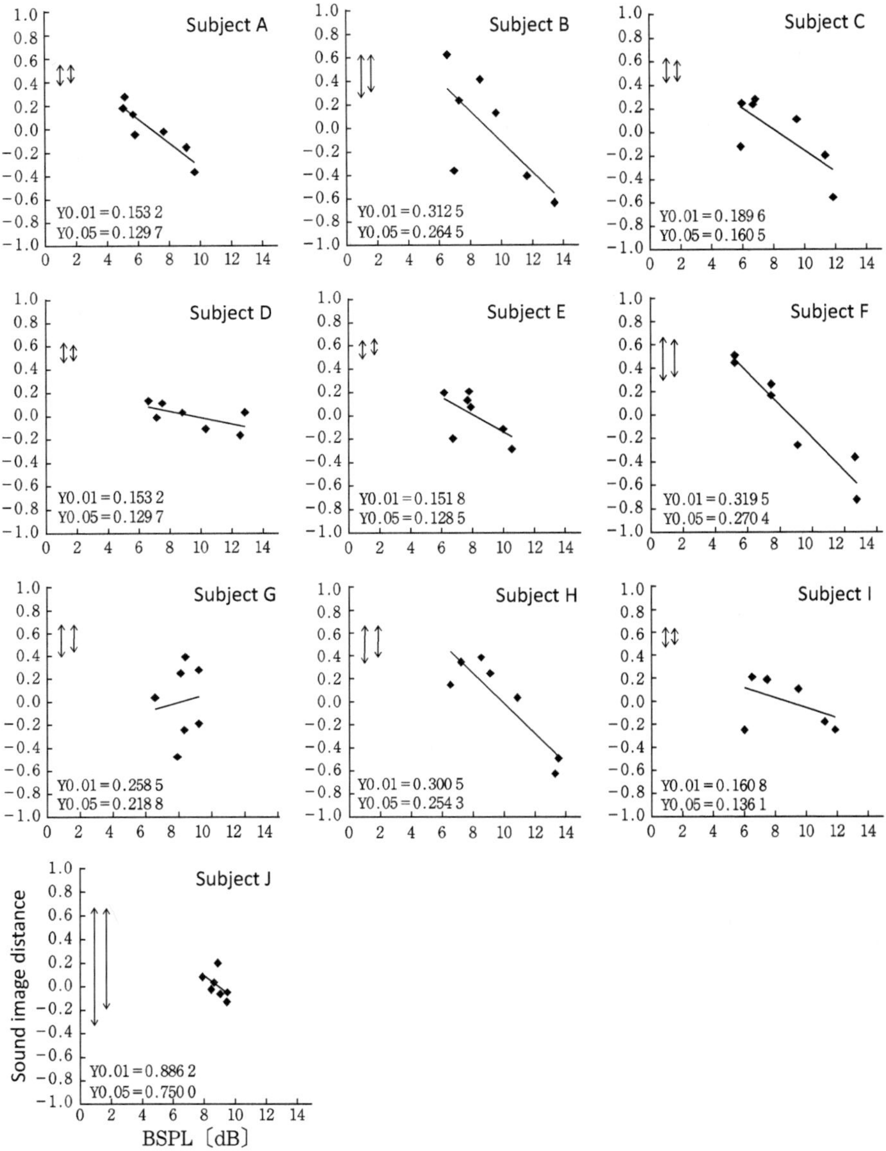

Fig. 7.15 Relationship between relative BSPL and sound image distance for each of ten subjects

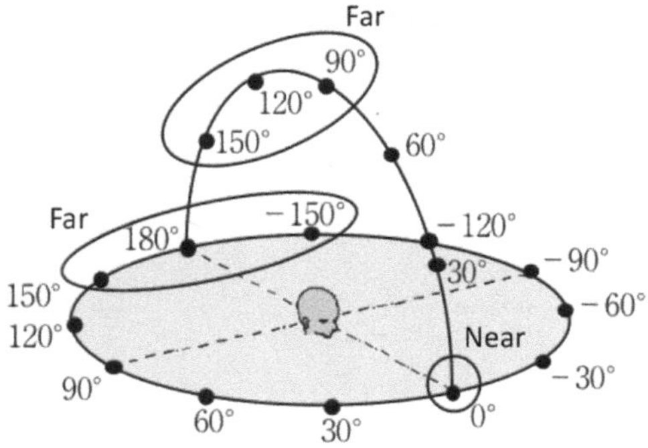

Fig. 7.16 Sound image distance in the horizontal plane and the median plane

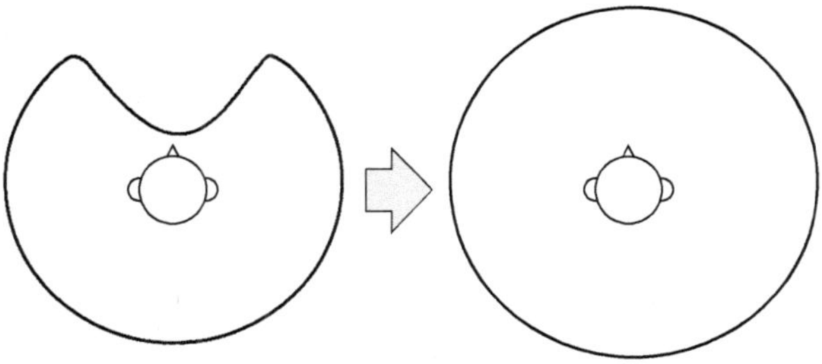

Fig. 7.17 Sound image distance in case of equal output sound pressure and equal BSPL

References

Gardner MB (1969) Distance estimation of 0°or apparent 0°–oriented speech signals in anechoic space. J Acoust Soc Am 45:47–53

Gotoh T, Kimura Y, Kurahashi A, Yamada A (1977) A consideration of distance perception in binaural hearing. J Acoust Soc Jpn (E) 33:667–671

Robinson DW, Whittle LS (1960) The loudness of directional sound field. Acust 10:74–80

Von Bekesy G (1949) The moon illusion and similar auditory phenomena. Am J Psychol 62:540–552

Chapter 8
Speech Intelligibility and HRTF

Abstract This chapter describes the effects of the HRTF on speech intelligibility. An HRTF causes an interaural phase difference. Under the presence of a masker, the threshold of the target sound is affected by the relationship in the interaural phase difference between the target sound and the masker. In other words, the threshold of the target sound is changed by the incident directions of the target sound and the masker. This also appears as a difference in speech intelligibility.

8.1 Binaural Masking Level Difference

When the target sound (maskee) and the interfering sound (masker) are presented to both ears through headphones, the masked threshold of the target sound, the sound pressure level at which the target sound can be heard in the presence of disturbing sound, changes according to the interaural phase difference (interaural time difference). In other words, the target sound becomes easier to hear or harder to hear.

Based on the masking threshold when both the masker and maskee are presented to a single ear (N_mS_m), the amount of change in the masking threshold when the masker and maskee are presented to both ears with a phase difference is referred to as the binaural masking level difference (BMLD). Here, the subscript m indicates monaural. The BMLD is known to be 12–15 dB at N_0S_π (interaural phase difference of the masker and the maskee are 0° and 180°, respectively) (Blauert 1996).

Moreover, the amount of change in speech intelligibility, not in the masking threshold, using a voice as maskee is referred to as the binaural intelligibility level difference (BILD).

Experiments were performed in which pink noise as a masker and click trains as a maskee were presented by loudspeakers placed on the horizontal plane or the median plane (Saberi et al. 1991). For a masker presented from the front, when the maskee was presented from just lateral direction (azimuth of ±90°), the masking threshold decreased by approximately 15 dB, as compared with the case in which the maskee was presented from the front. On the other hand, when the maskee was presented from the rear (azimuth of 180°), the masking threshold was approximately the same

K. Iida, *Head-Related Transfer Function and Acoustic Virtual Reality*,
https://doi.org/10.1007/978-981-13-9745-5_8

as that presented from the front. When the maskee was presented from above (vertical angle of 60°–150°), the masking threshold decreased by approximately 8 dB.

8.2 Influence of Incident Direction on Word Intelligibility

As described above, the interaural phase difference caused by the HRTF changes the masking threshold, therefore, it is assumed that the speech intelligibility changes depending on the incident direction of the target sound and the interference sound when the target sound is speech. In order to verify this finding, the following experiment was conducted.

In the anechoic chamber, the preceding sound was presented from the front, and a single echo was presented from either the right horizontal plane (in 30° steps) or the upper median plane (in 30° steps), as shown in Fig. 8.1. The time delay and sound pressure level of the single echo compared with the preceding sound were 1s and 0 dB, respectively.

The sound source signal was a quadruple word in which four-mora words are connected four by one at intervals of 1s, as shown in Fig. 8.2. The first word and the fourth word had a timing such that the words could be heard alone, but the second word and the third word always overlapped with another word in time. In this case, the preceding sound and the single echo can be both the target sound and the interference sound.

The subjects were nine students in their twenties (five males and four females).

The intelligibility for all words from the first word to the fourth word is shown in Fig. 8.3. Figure 8.3(a) shows the case in which a single echo comes from the horizontal plane, and Fig. 8.3(b) shows the case of the median plane.

When single echo comes from the horizontal plane, the intelligibility is high at an azimuth of approximately 90° and is low at the front and rear. This is the same tendency as the masking threshold. The results of a chi-square test showed statistically significant differences between 0° and 60° and between 0° and 90°.

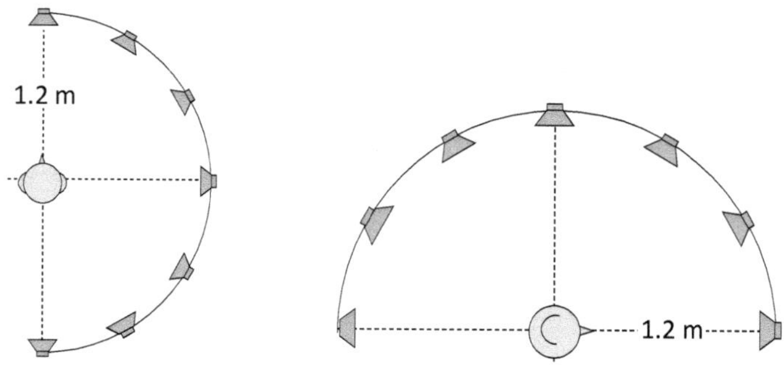

Fig. 8.1 Loudspeaker arrangement for a single echo

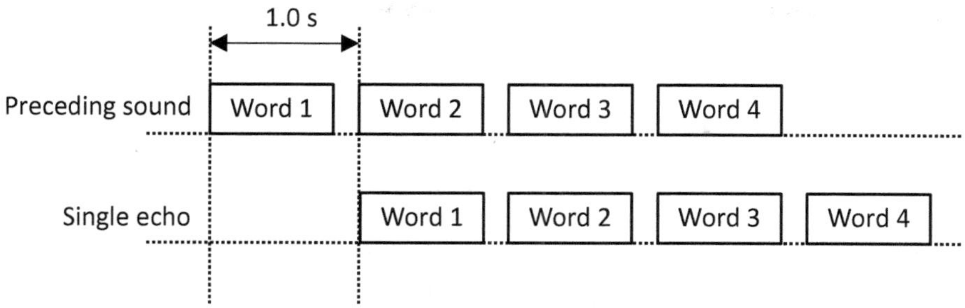

Fig. 8.2 Temporal structure of the speech stimuli

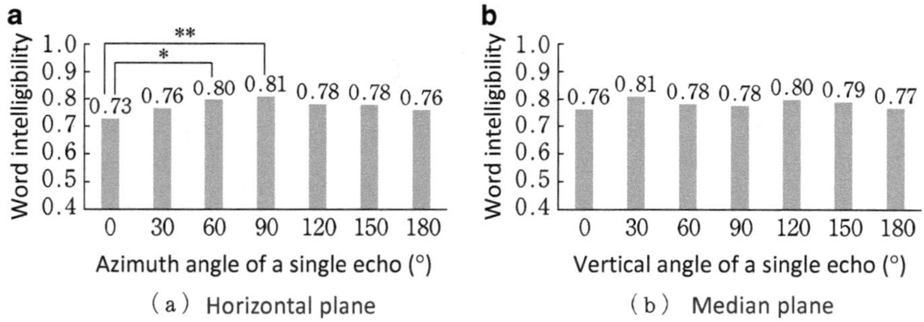

Fig. 8.3 Word intelligibility for each incident angle of a single echo. (**a**) horizontal plane and (**b**) median plane. $p < 0.05$, $p < 0.01$

On the other hand, when a single echo comes from the median plane, the effect of the incident rise angle is hardly noticeable.

Furthermore, word intelligibility was obtained for each order of word presentation, as shown in Figs. 8.4 and 8.5. The intelligibility of the first and fourth words, which can be heard alone, was high in both the horizontal and median planes, and the difference due to the incident direction was small.

In the horizontal plane, the intelligibility of the second and third words was higher at the side, as compared to the front. A significant difference was found for the second word.

In the median plane, there is a high intelligibility for an upward angle, as compared to the front. However, there is no commonality between the second and third words.

The above results suggest that the word intelligibility is influenced by the incident direction in the presence of a masker, the influence of which is qualitatively consistent with the characteristics of the masking threshold.

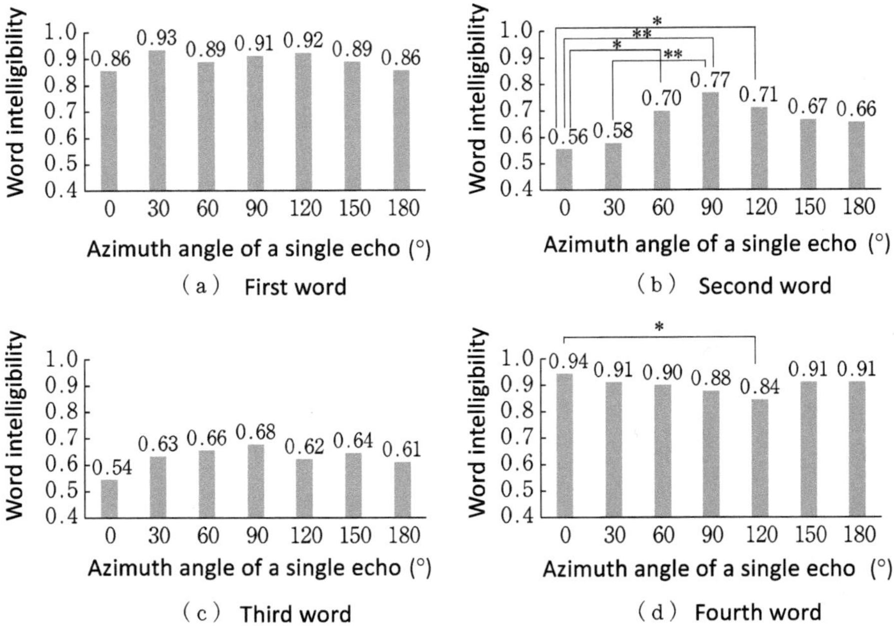

Fig. 8.4 Word intelligibility for each azimuth angle of a single echo. $*$ $p < 0.05$, $**$ $p < 0.01$

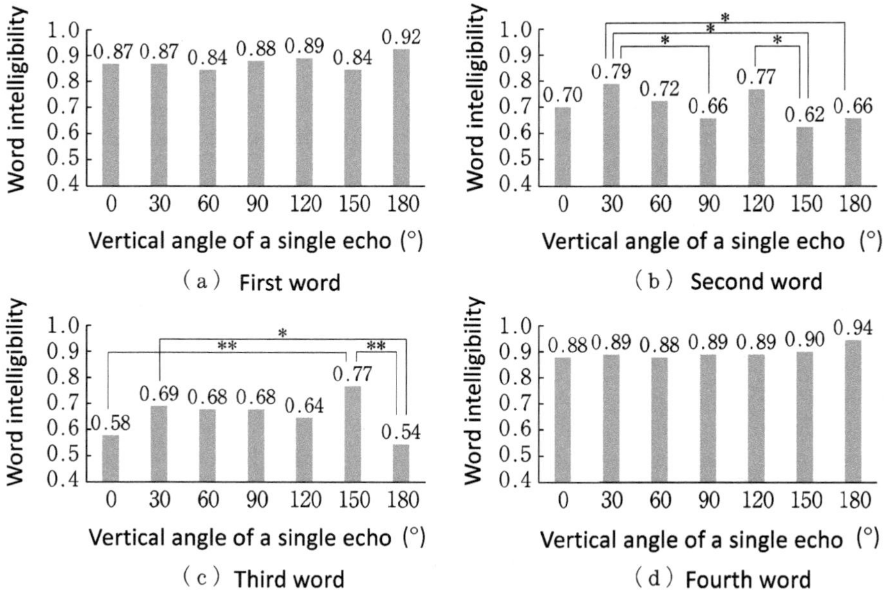

Fig. 8.5 Word intelligibility for each vertical angle of a single echo. $*$ $p < 0.05$, $**$ $p < 0.01$

References

Blauert J (1996) Spatial hearing–the psychology of human sound localization, Rev edn. The MIT Press, Cambridge, MA, pp 257–271

Saberi K, Dostal L, Sadralodabai T, Bull V, Perrott DR (1991) Free–field release from masking. J Acoust Soc Am 90:1355–1370

Chapter 9
Measurement Method for HRTF

Abstract In order to actually measure HRTFs, practical knowledge is also required in addition to understanding the theory of the HRTF. This chapter introduces the author's knowledge and experience as far as possible, so as to enable the readers to measure HRTFs by themselves.

9.1 Configuration of the Measurement System

An example of a HRTF measurement system is shown in Fig. 9.1. The digital sound signal for measurement sent from the PC is converted to an analog sound wave by an audio interface, amplified by an amplifier, and emitted from a loudspeaker installed in an anechoic chamber. The sound wave is picked up by earplug-type microphones attached to both of the subject's ears, amplified by a microphone amplifier, converted to a digital signal by an audio interface, and recorded on a PC.

Figure 9.2 shows a photograph of measurement of the HRTFs in the upper median plane.

Each component of the measurement system will be described in detail in the following sections.

9.2 Measurement Signal

As a sound source for measuring HRTFs, M-sequence signals and swept-sine signals are widely used. In particular, since the swept-sine signal has a long duration and a large energy as compared with a single pulse, an impulse response can be measured with a high signal-to-noise ratio.

M-sequence signals can also measure an impulse response with a high signal-to-noise ratio, but these signals are not suitable for measurement in a space with air flow because they are susceptible to fluctuations in the sound field.

K. Iida, *Head-Related Transfer Function and Acoustic Virtual Reality*,
https://doi.org/10.1007/978-981-13-9745-5_9

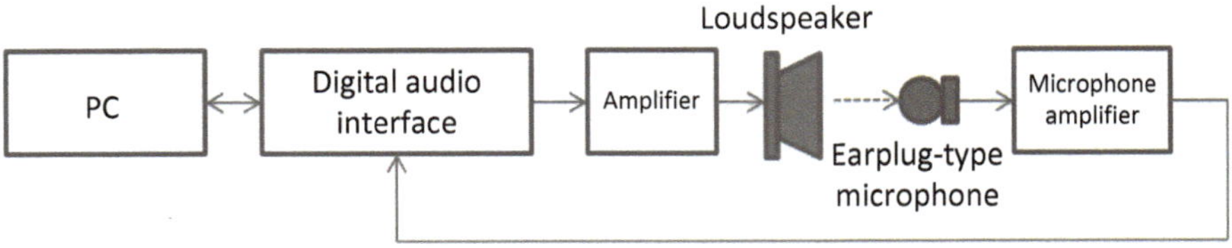

Fig. 9.1 Configuration of HRTF measurement system

Fig. 9.2 Photograph of measurement of the HRTFs in the upper median plane

The swept-sine signal is created by increasing the phase of the Fourier transform of a single pulse in proportion to the square of the frequency, as shown in Eq. (9.1), and applying an inverse Fourier transform to this signal:

$$S(k) = \begin{cases} e^{\frac{-j\pi k^2}{N}} & 0 \leqq k \leqq \dfrac{N}{2} \\ S^*(N-k) & \dfrac{N}{2} < k < N \end{cases} \tag{9.1}$$

where $*$ denotes a conjugate complex, and N is a power of two.

An impulse response can be obtained by inputting this signal into the system to be measured and convoluting the reverse swept-sine signal represented by Eq. (9.2) into the obtained output:

```
clear;
n=15;  N=2^n;                    // Length of signal
scale=10000;                     // Maximum amplitude
flag=1;            //Swept-sine signal

S=zeros(1,N);

for k=0:N/2;
      kk=k+1;
S(kk)=cos(%pi*k*k/N+0.5*%pi*k)-sin(%pi*k*k/N+0.5*%pi*k)*%i*flag;
end

for k=N/2+1:N-1;
      kk=k+1;
      S(kk)=conj(S(N-kk+2));
end

s=ifft(S);
s=s/max(real(s))*scale;

clf
subplot(211);plot2d(s)
xlabel(' Sample');
ylabel(' Amplitude');
square(0,-scale,N,scale)
subplot(212);plot2d(s)
xlabel(' Sample');
ylabel(' Amplitude');
square(N/4-300,-scale,N/4+2000,scale)
```

Fig. 9.3 Sample program for swept-sine signal generation

$$S^{-1}(k) = \begin{cases} e^{\frac{j\pi k^2}{N}} & 0 \leqq k \leqq \frac{N}{2} \\ S^*(N-k) & \frac{N}{2} < k < N \end{cases} \tag{9.2}$$

Sample programs (Scilab) for creating a swept-sine signal and the corresponding reverse swept-sine signal are shown in Figs. 9.3 and 9.4. Furthermore, waveforms of the swept-sine signal and the corresponding reverse swept-sine signal are shown in Figs. 9.5 and 9.6.

```
clear;
n=15;   N=2^n;                    //Length of signal
scale=10000;                      //Maximum amplitude
flag=-1;              // Swept-sine signal

S=zeros(1,N);

for k=0:N/2;
      kk=k+1;
S(kk)=cos(%pi*k*k/N+0.5*%pi*k)-sin(%pi*k*k/N+0.5*%pi*k)*%i*flag;
end

for k=N/2+1:N-1;
      kk=k+1;
      S(kk)=conj(S(N-kk+2));
end

s=ifft(S);
s=s/max(real(s))*scale;

clf
subplot(211);plot2d(s)
xlabel('Sample');
ylabel('Amplitude');
square(0,-scale,N,scale)
subplot(212);plot2d(s)
xlabel('Sample');
ylabel('Amplitude');
square(N/4*3-2000,-scale,N/4*3+200,scale)
```

Fig. 9.4 Sample program for reverse swept-sine signal generation

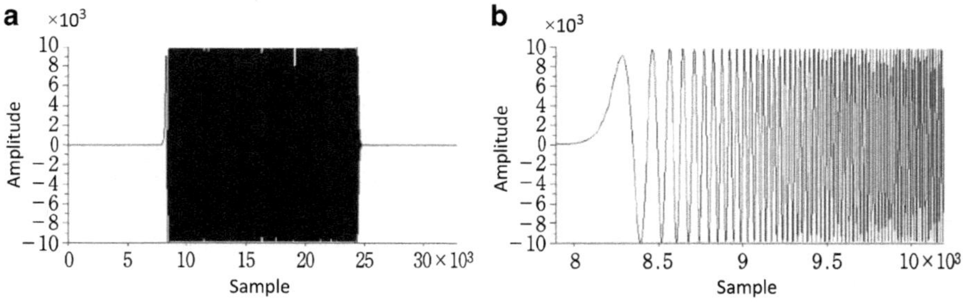

Fig. 9.5 Waveform of swept-sine signal. (**a**) Overall view, (**b**) Enlarged view of rising part

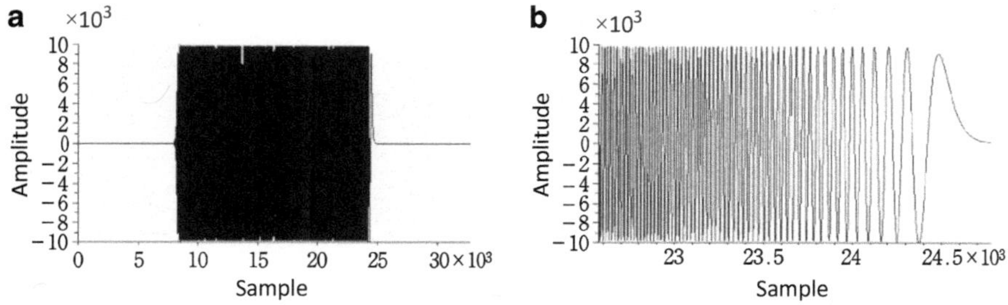

Fig. 9.6 Waveform of reverse swept-sine signal. (**a**) Overall view, (**b**) Enlarged view of decay part

Fig. 9.7 Loudspeaker for HRTF measurement (FE83E in SV-70 enclosure)

9.3 Loudspeakers

The loudspeaker is required to have a single acoustic center and to have continuity in phase frequency characteristics. Therefore, a single-cone-type loudspeaker is desirable. In order to minimize reflection from the cabinet (enclosure), the cabinet should be as small as possible or have a circular cross section.

The author used a FE83E (Fostex) loudspeaker, the diameter of which is 80 mm, and a SV-70 (Daito voice) as a cabinet (Fig. 9.7).

9.4 Microphones

A small microphone unit is used to pick up sound at the entrance of each ear canal. The author has long used a WM64AT102 (Panasonic) with a diameter of 5 mm, but production appears to have been discontinued. An acceptable replacement microphone is the FG3329 (Knowles). A probe microphone (4182 (B&K), ER-7C (Etymotic Research)) may also be used. However, for a probe microphone, it is

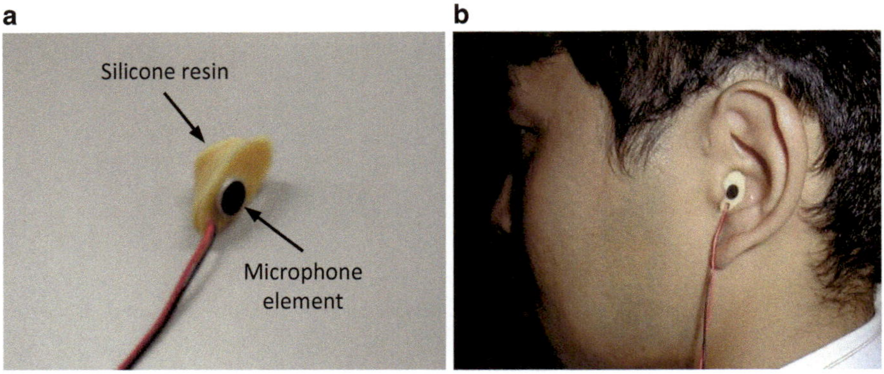

Fig. 9.8 Photographs of (**a**) earplug-type microphone and (**b**) its placement in ear canal of subject

necessary to pay close attention to reproducibility of the installation position at the entrance of the ear canal.

Figure 9.8(a) shows the earplug-type microphone fabricated using the WM64AT102, and Fig. 9.8(b) shows its placement in the ear canal of a subject. The method for fabricating the earplug-type microphone is described in detail in Appendix A.7.

9.5 Subjects

The following are notes related to the subjects. First, care must be taken to maintain the subject's posture during measurement. When tired, the subject has a tendency to look down. In such a case, the rising angle of the sound source becomes larger than the setting. It is necessary to stress to the subject the importance of facing forward. Attaching a sticker to the point to be watched helps the subject to maintain his/her head orientation constant.

There is also concern about the influence of differences in the subject's hair style on the HRTFs. Changes in HRTFs of up to 3 kHz caused by adding hair to a sphere head model have been reported (Treeby et al. 2007). According to this report, by adding hair, the ITD increases by 20–25 μs at the lateral direction. However, this is considered to have no effect on the sense of direction because it is smaller than the JND (72 μs) of the ITD at the lateral direction.

By adding hair, the ILD increases by approximately 4 dB at 3 kHz for the lateral direction and exceeds the JND (1 dB). However, for most sound-source directions and frequencies, the change in the ILD is less than the JND.

In the measurement using the KEMAR dummy head, the spectral notch around 10 kHz is reported to be slightly shallower by adding hair. However, there is almost no effect at lower frequencies (Burkhard and Sachs 1975). Although the level of the notch changes somewhat by the addition of hair, the notch frequency is unaffected.

The above suggests that, although there are some physical changes in the HRTF due to the increase or decrease of hair, the volume of hair does not affect the perception of direction of a sound image.

9.6 Derivation Method for HRTFs

As described in Chap. 1, the HRTF can be obtained as follows:

$$H_{\mathrm{l,r}}(s,\alpha,\beta,r,\omega)=\frac{\mathcal{F}\left[g_{\mathrm{l,r}}(s,\alpha,\beta,r,t)\right]}{\mathcal{F}[f(\alpha,\beta,r,t)]} \tag{9.3}$$

where $g_{l,r}$ is the impulse response between a sound source and the entrance of the ear canal of the subject in the free field, f is the impulse response between a sound source and the point corresponding to the center of the subject's head in the free field without the subject, and $\mathcal{F}$ denotes the Fourier transform.

In the numerator of Eq. (9.3), $g_{\mathrm{l,r}}$ is obtained by convolving reverse swept-sine signals to swept-sine signals recorded at earplug-type microphones attached to the entrances of the ear canals.

Similarly, in the denominator of Eq. (9.3), f is obtained by convolving reverse swept-sine signals to swept-sine signals recorded at each earplug-type microphone placed at a position corresponding to the center of the subject's head in the absence of the subject. This is the impulse response of the measurement system itself.

HRTFs are obtained by performing complex division on the results of the Fourier transforms of $g_{\mathrm{l,r}}$ and f.

However, care must be taken here. Even for measurement in an anechoic chamber, the relative energy of the noise component increases as the impulse response is increased. Since HRIRs converge in several milliseconds, it is wise to extract only that part using the time window.

The author obtains HRTFs by the following steps.

(1) The sample number at which the impulse response of the front direction, $g_{\mathrm{l,r}}$, takes the maximum absolute value of the amplitude is obtained.
(2) The sample number, which is 50 samples before the sample obtained in (1), is used as the start point of the time window of $g_{\mathrm{l,r}}$ and f in all directions. By going back 50 samples, it is possible to extract from the first part of the response even in the lateral direction, where the sound reaches the ipsilateral ear earliest (here, we assume 48-kHz sampling).
(3) The sample that zero-crosses first after more than 128 samples after the start sample is set as the end point of the time window. Since amplitudes are close to zero at both the start and end points, a rectangular window is used.
(4) Each $g_{\mathrm{l,r}}$ and f cut out by the time window is zero-padded to 512 samples.
(5) An FFT for 512 samples is performed, and the HRTF is obtained by complex division. The frequency resolution in this case is 93.75 Hz.

Fig. 9.9 Photographs of (**a**) small loudspeaker and (**b**) its placement in ear canal of subject. (Zotkin et al. 2006)

9.7 Short-Time HRTF Measurement Method

Although development of the measurement signal described in Sect. 9.2 has made it possible to measure HRTFs for a given direction at a high SN ratio and in a short time. However, measurement of HRTFs for many directions still requires a long time.

In order to reduce the total measurement time, a continuous measurement method, which measures HRTFs while rotating the subject at a constant speed, has been developed.

Moreover, a high-speed measurement method of HRTFs based on the reciprocity law has also been studied. The reciprocity law states that the sound pressure observed at point B caused by a sound source at point A is equal to the sound pressure observed at point A when the sound source is placed at point B.

Based on the reciprocity law, the HRTFs for multiple directions can be measured at the same time by inserting a small loudspeaker, as shown in Fig. 9.9(a), in the entrance of the ear canal, as shown in Fig. 9.9(b), and installing microphones in multiple directions. The results of the measurement of the HRTFs of the KEMAR dummy head were obtained using the reciprocity law. The results showed that the outline of the amplitude spectrum of the HRTFs was similar to that measured using the conventional method, although they were not exactly identical (Zotkin et al. 2006). There remain problems to be solved, such as improvement of the SN ratio, for practical use.

References

Burkhard MD, Sachs RM (1975) Anthropometric manikin for acoustic researches. J Acoust Soc Am 58:214–222

Treeby BE, Pan J, Paurobally RM (2007) The effect of hair on auditory localization cues. J Acoust Soc Am 122:3586–3597

Zotkin DN, Duraiswami R, Grassi E, Gumerov NA (2006) Fast head-related transfer function measurement via reciprocity. J Acoust Soc Am 120:2202–2215

Chapter 10
Signal Processing of HRTF

Abstract In order to analyze HRTFs and apply them to acoustic VR systems, signal processing technique is required. This chapter introduces the methods for calculating the ITD and ILD, and extracting the spectral cues. Furthermore, the methods for convolution of the HRIRs and a sound source signal is introduced.

10.1 Method for Calculating the ITD and the ILD

An example method for calculating the ITD from the HRTF is described below.

(1) Apply a minimum phase low-pass filter with a cut-off frequency of 1.6 kHz to HRIR (512 samples). As described in Sect. 2.3.1, the time difference of the waveform of the binaural input signals acts as a cue for the perception of lateral angle only for the components below 1600 Hz.
(2) In order to improve the time resolution of the ITD, increase the sampling frequency by eight times (e.g. 48 kHz $\times$ 8 = 384 kHz). As a result, the time resolution is approximately 2.6 μs, and the angle resolution is approximately 0.3°.
(3) The time difference, τ, at which the interaural cross correlation function, Φ (Eq. 10.1), becomes a maximum is taken as the ITD:

$$\Phi_{1,\mathrm{r}}(\tau) = \lim_{T\to\infty} \frac{\int_{-T}^{T} HRIR_1(t) \times HRIR_\mathrm{r}(t-\tau)dt}{\sqrt{\int_{-T}^{T} HRIR_1^2(t) \times HRIR_\mathrm{r}^2(t)}} \tag{10.1}$$

where $|\tau| \le 1000$ μs. The subscripts l and r indicate the left and right ears, respectively.

The ILD for each frequency band is calculated as follows.

(1) Apply a 1/3 octave or 1/1 octave bandpass filter to the HRIR, the length of which is 512 samples.

K. Iida, *Head-Related Transfer Function and Acoustic Virtual Reality*,
https://doi.org/10.1007/978-981-13-9745-5_10

(2) Obtain the root mean square of the HRIR of each band, and calculate the level difference between the left and right ears.

Or,

(1) the HRIR of 512 samples is zero-padded to 48,000 samples, and the frequency resolution is thus 1 Hz.
(2) The band level of each band is obtained by FFT, and the level difference between the left and right ears is calculated.

10.2 Extracting Method of Spectral Cues

The method to obtain the amplitude spectrum of the HRTF from the HRIR and extract the spectral cue is as follows (Iida et al. 2014).

(1) Detect the sample for which the absolute amplitude of the HRIR (512 samples) is a maximum.
(2) Clip the HRIR using a four-term, 96-point Blackman-Harris window (Fig. 10.1, 2 N = 96), adjusting the temporal center of the window to the maximum sample detected in (1).
(3) Prepare a 512-point array, all of the values of which are set to zero, and overwrite the clipped HRIR in the array, where the maximum sample of the clipped HRIR should be placed at the 257th point in the array.
(4) Obtain the amplitude spectrum of the 512-point array by FFT. Then, find the local maxima and local minima of the amplitude using the difference method, which replaces the derivative with the finite difference.
(5) Define the lowest two frequencies of the local maxima above 3 kHz as P1 and P2, and define the lowest two frequencies of the local minima above P1 as N1 and N2.

The reason why the early part of the HRIR is used is described below.

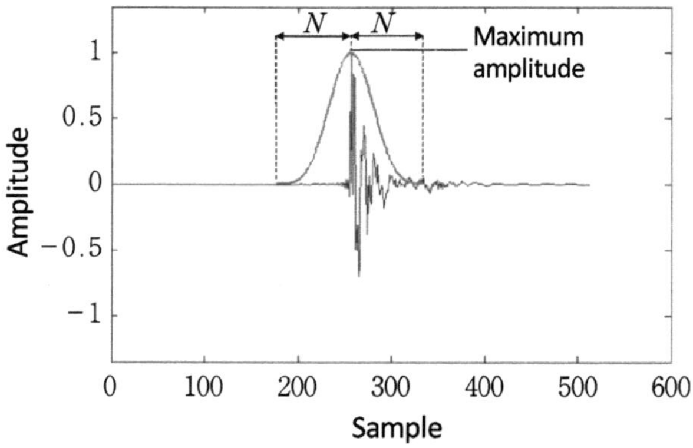

Fig. 10.1 HRIR and four-term 96-point Blackman-Harris window

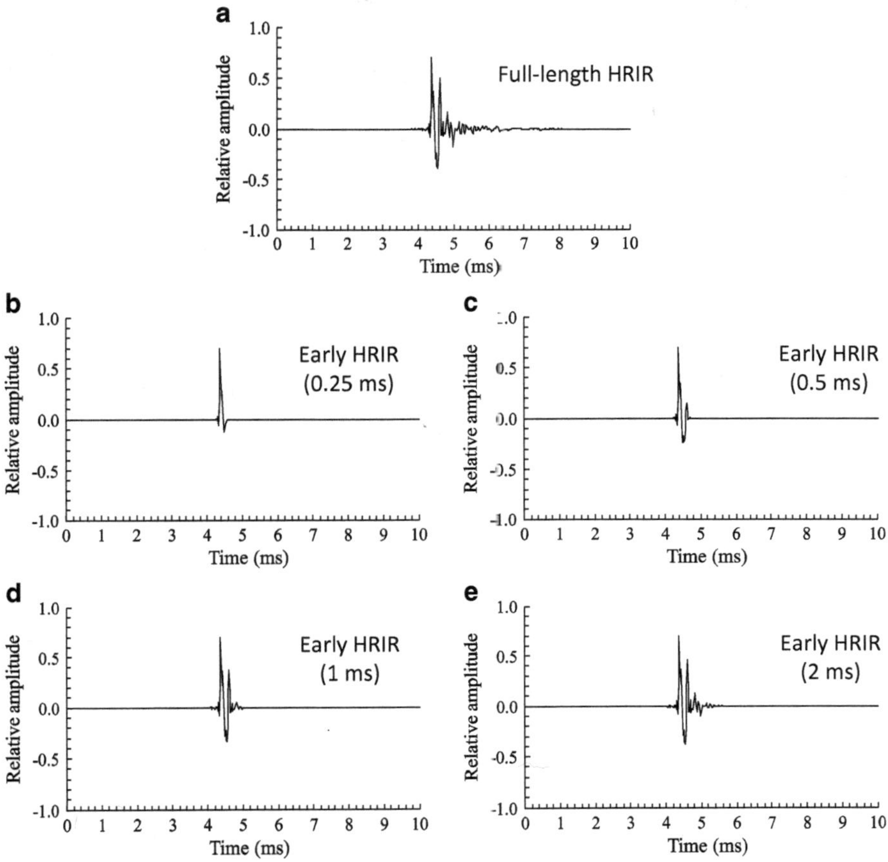

Fig. 10.2 Full-length HRIR and early HRIRs for front direction. (**a**) full-length HRIR, (**b**) early HRIR (0.25 ms), (**c**) early HRIR (0.5 ms), (**d**) early HRIR (1 ms), and (**e**) early HRIR (2 ms). (Iida and Oota 2018)

The notches and peaks are reported to be generated in the pinna. When the cavities of the pinna are occluded by clay, the notches and peaks vanish (Iida et al. 1998), and the front-back confusion of a sound image increases (Gardner and Gardner 1973, Musicant and Butler 1984, Iida et al. 1998). At the notch frequency, the anti-nodes are generated at the cymba of concha and the triangular fossa, and a node is generated at the cavity of concha (Takemoto et al. 2012). Peaks are considered to be generated by the resonances of the pinna (Shaw and Teranishi 1968).

The effect of the pinnae is considered to be included in the early part of the HRIR because the response from the pinna arrives at the receiving point (the entrance of the ear canal) earlier than that from the torso. Therefore, information on the outline of N1, N2, P1, and P2 is supposed to be included in the early part of HRIR (hereinafter referred to as the early HRIR).

Figure 10.2 shows examples of a full-length HRIR and the early HRIRs of a subject for the front direction (Iida and Oota 2018). The response of the full-length

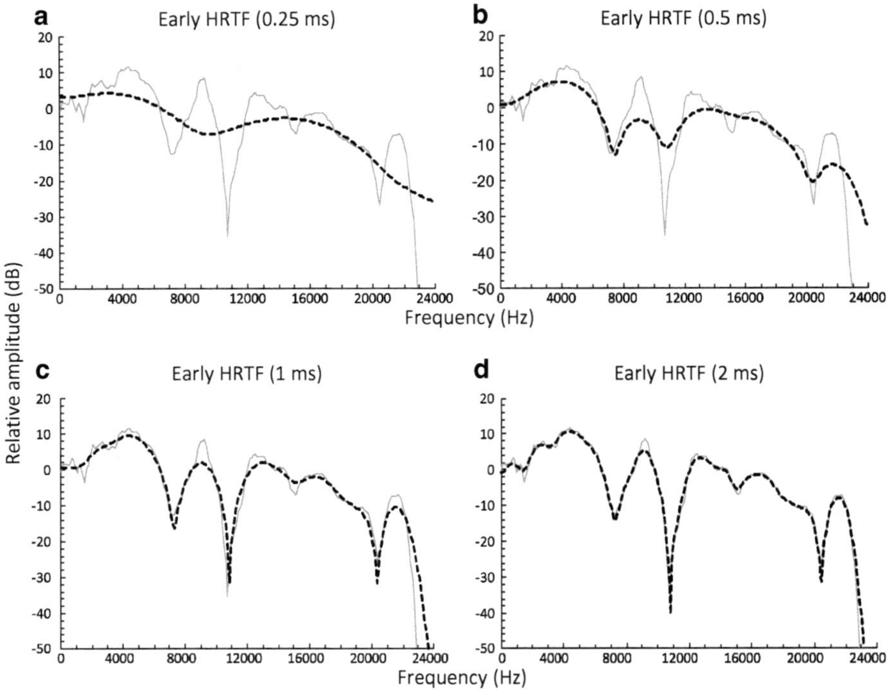

Fig. 10.3 Amplitude spectra of full-length HRTF and early HRTFs for front direction. The solid and broken lines indicate the full-length HRTFs and the early HRTFs, respectively. (**a**) early HRTF (0.25 ms), (**b**) early HRTF (0.5 ms), (**c**) early HRTF (1 ms), and (**d**) early HRTF (2 ms). (Iida and Oota 2018)

HRIR converged within 4 ms from the maximum sample. The early HRIR of 0.25 ms includes only the positive part of the first response. The early HRIR of 0.5 ms includes the positive and negative parts of the first response. However, the absolute value of the negative part was decreased by the temporal window. The early HRIR of 1 ms (2 N = 96) includes the response until the positive part of the second response, which was, however, decreased by the temporal window. The early HRIR of 2 ms includes most of the responses, except for the fine responses of the later part.

Figure 10.3 shows the amplitude spectra of the full-length HRTFs and early HRTFs of the same subject for the front direction. The amplitude spectra were obtained by the FFT with 512 samples. The full-length HRTFs (solid lines) include several notches and peaks.

For the early HRTF of 0.25 ms, one notch and two peaks were observed. The frequencies of the notch and peaks did not coincide with those of the full-length HRTFs. The notch was shallow, and the peak level was low.

For the early HRTF of 0.5 ms, three notches were observed. The frequencies of the notches were approximately the same as those of the full-length HRTF (the frequency difference was within 281.75 Hz). However, the levels of the notches did not coincide with those of the full-length HRTF. Four peaks were observed, and the

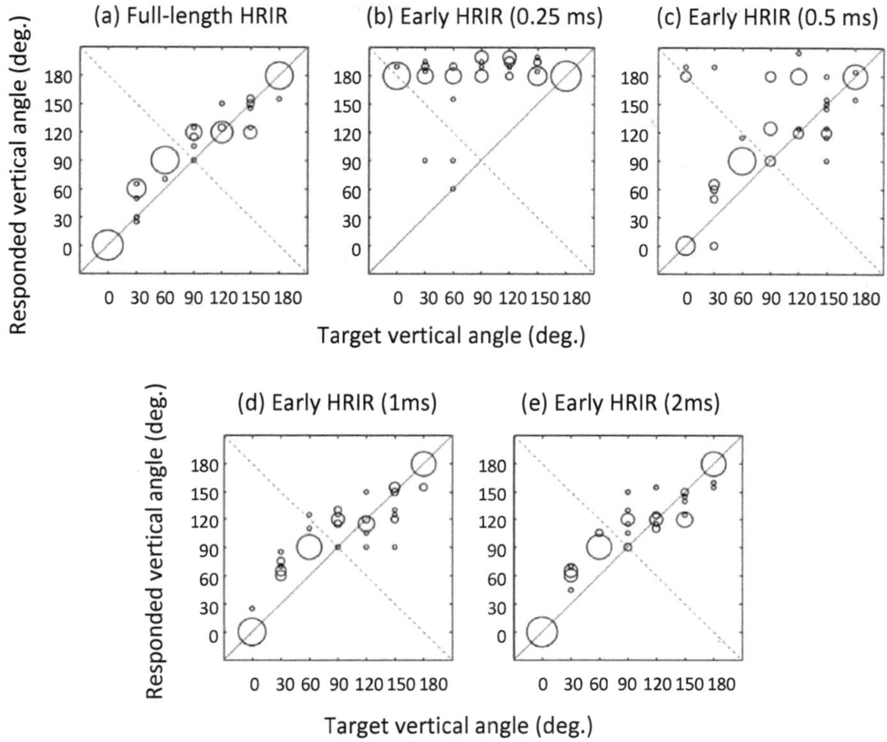

Fig. 10.4 Responses to full-length HRTFs and early HRTFs for seven vertical angles in upper median plane. (Iida and Oota 2018)

lowest-frequency peak coincided with that of the full-length HRTF. However, most of the other peaks did not coincide with those of the full-length HRTF.

For the early HRTF of 1 ms (2 N = 96), the outline of the notches and peaks was approximately the same as that of the full-length HRTF, while the fine structure differed from that of the full-length HRTF.

For the early HRTF of 2 ms, the amplitude spectrum coincided with that of the full-length HRTF to the last detail.

Figure 10.4 shows the results of localization tests using the subject's own full-length HRIRs and early HRIRs for the upper median plane. For the full-length HRIR, most of the responses were distributed around the target vertical angles. For the early HRIR of 0.25 ms, the responses were distributed rearward. For the early HRIR of 0.5 ms, the responses were around both the target vertical angle and 180° at target vertical angles of 0°, 90°, and 120°. On the other hand, for the early HRIRs of 1 and 2 ms, the distribution of the responses was approximately the same as that for the full-length HRIR for each target vertical angle.

These results suggest that the early HRIR of 1 ms includes information on the outline of the spectral notches and peaks with respect to the physical aspect and provides approximately the same vertical angle and distance of a sound image as the full-length HRIR in the upper median plane with respect to the perceptual aspect.

10.3 Method for Convolution of the HRIR and Sound Source Signal

10.3.1 Calculation in the Time Domain

In order to control the spatial characteristics of an arbitrary sound source signal by an acoustic VR system, convolution of the sound source signal and HRIRs or binaural room impulse responses (BRIRs) is required (see Appendix A2). Furthermore, this process is required to be performed in real time in many applications.

Assuming the impulse response of a certain system is h(t), the output signal y(t) when a signal x(t) is input into this system is expressed by convolution integrals as follows:

$$y(t) = x(t) * h(t) = \int_{-\infty}^{\infty} x(\tau)h(t - \tau)d\tau \tag{10.2}$$

This equation indicates that the output signal $y(t)$ at a certain time t is the sum of the product $x(\tau)h(t - \tau)$ with regard to τ.

Figure 10.5(a) shows an example of the impulse response $h(t)$. Here, $h(t)$ is assumed to be a response that exponentially decays with time. Figure 10.5(b)

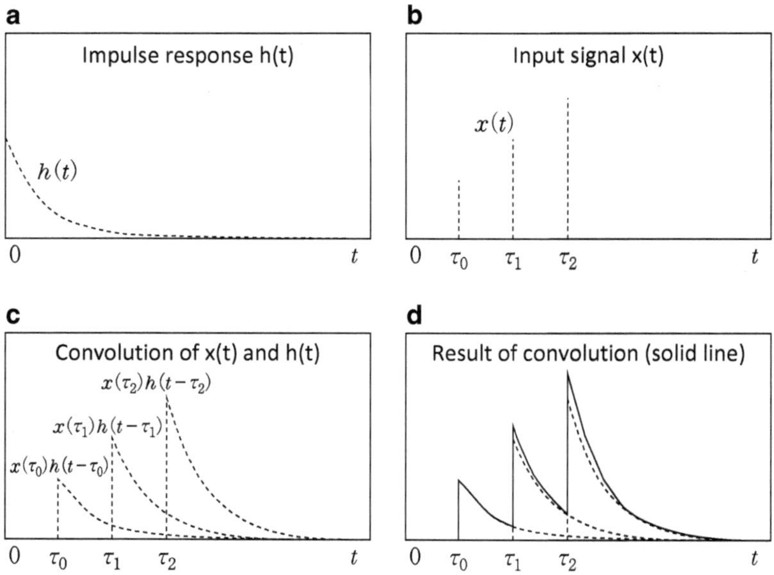

Fig. 10.5 Conceptual diagram of convolution x(τ)h(t -τ). (**a**) an impulse response h(t), (**b**) three discrete input signals, (**c**) convolution of each of the three discrete input signals and an impulse response, and (**d**) final output signal y(t)

shows an example of an input signal, where we assume that there is an input only at $t = \tau_0, \tau_1, \tau_2$. Figure 10.5(c) shows the convolution of each of the three discrete input signals and an impulse response.

The first exponential decay curve represents the response $x(\tau_0)h(t - \tau_0)$ to the input signal $x(\tau_0)$ as a function of time t. Similarly, the second and third exponential decay curves represent the responses $x(\tau_1)h(t - \tau_1)$ and $x(\tau_2)h(t - \tau_2)$ to the input signals $x(\tau_1)$ and $x(\tau_2)$, respectively. The final output signal y(t) is expressed as the sum of $x(\tau_0)h(t - \tau_0)$, $x(\tau_1)h(t - \tau_1)$, and $x(\tau_2)h(t - \tau_2)$, as shown in Fig. 10.5(d).

In this way, the output signal $y(t)$ when the signal $x(t)$ is input into a system having impulse response $h(t)$ is expressed as the sum of all $x(\tau)h(t - \tau)$, which is reached at time t.

The sample program for convolution (Scilab) is shown in Fig. 10.6. Figure 10.7 (a) is the waveform of a music signal. Figure 10.7(b) shows the impulse response from a sound source to a receiving point in a room. Assuming the time when an impulse is emitted from the sound source to be t = 0, direct sound arrives at the receiving point in approximately 30 ms. Then, a large number of reflections arrive,

```
clear
stacksize('max');
fs = 48000;      // Sampling frequency
T = 1/fs;        // Sampling period

// -----   Read sound source   -----//
x = wavread('music.wav');
len_x = length(x);

// -----   Read impulse response  ----- //
fid=mopen('ImpulseResponse.bin','rb');
h = mget(fs*10,'f',fid);
mclose(fid)
len_h = length(h);

// ----- Convolution ----- //
y = convol(x,h);
len_y = length(y);

// ----- Plot ----- //
clf
subplot (3,1,1); plot2d([0:len_x-1]*T,x)
xlabel ('Time (s)') ; ylabel('amplitude);'
subplot (3,1,2); plot2d([0:len_h-1]*T,h);
xlabel ('Time (s)') ; ylabel('amplitude);'
subplot (3,1,3); plot2d([0:len_y-1]*T,y);
xlabel ('Time (s)') ; ylabel('amplitude);'
```

Fig. 10.6 Sample program for convolution

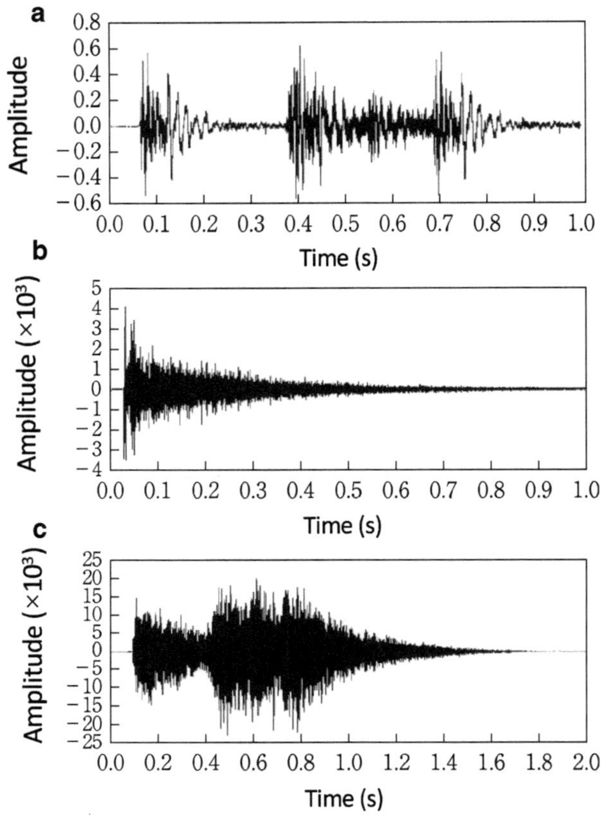

Fig. 10.7 (**a**) Waveform of sound source (music), (**b**) impulse response of room, and (**c**) result of convolution

and the response converges in approximately 1 s. When this music signal is emitted in this sound field, the signal at the receiving point is as shown in Fig. 10.7(c).

A sample program for convolution of a sound source signal and impulse responses from the sound source to the entrances to a listener's ears are shown in Fig. 10.8. The results of the convolution are shown in Fig. 10.9.

10.3.2 Calculation in the Frequency Domain

A. *Method based on FFT*

Assuming the length of the sound source signal $x(t)$ to be M and that of the impulse response $h(t)$ to be N, the convolution integral of Eq. (10.2) requires $M \times N$ product-sum operations.

```
clear
fs = 48000;      // Sampling frequency
T = 1/fs         // Sampling period

// -----  Read sound source   -----//
x = wavread('music.wav');
len_x = length(x);

// -----     Read impulse responses        ----- //
fid1=mopen('ImpulseResponseL.bin','rb');
fid2=mopen('ImpulseResponseR.bin','rb');
hL = mget(fs*10,'f',fid1);
hR = mget(fs*10,'f',fid2);
mclose(fid1)
mclose(fid2)
len_hL = length(hL);
len_hR = length(hR);

// -----Convolution   ----- //
yL = convol(x,hL);
yR = convol(x,hR);
len_yL = length(yL);
len_yR = length(yR);

// ----- Plot ----- //
clf
subplot (311); plot2d([0:len_x-1]*T,x);
xlabel ('Time (s)') ; ylabel('amplitude);'
square(0,-1,len_x*T,1);
subplot (323); plot2d([0:len_hL-1]*T,hL);
xlabel ('Time (s)') ; ylabel('amplitude);'
square(0,-10000,len_hL*T,10000);
subplot (324); plot2d([0:len_hR-1]*T,hR);
xlabel ('Time (s)') ; ylabel('amplitude);'
square(0,-10000,len_hR*T,10000);
subplot (325); plot2d([0:len_yL-1]*T,yL);
xlabel ('Time (s)') ; ylabel('amplitude);'
square(0,-50000,len_yL*T,50000);
subplot (326); plot2d([0:len_yR-1]*T,yR);
xlabel ('Time (s)') ; ylabel('amplitude);'
square(0,-50000,len_yR*T,50000);
```

Fig. 10.8 Sample program for convolution for ear-input signal

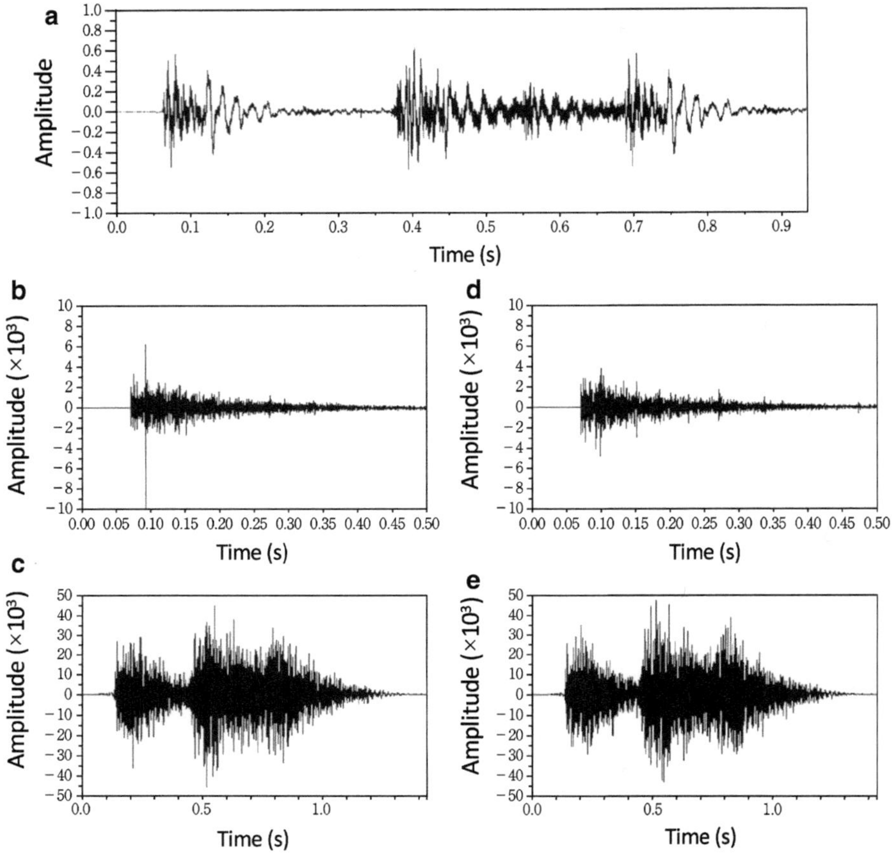

Fig. 10.9 (**a**) Waveform of sound source (music), (**b**) and (**c**) impulse responses at entrances of listener's ears, and (**d**) and (**e**) results of convolution

The length of a binaural impulse response in a typical room (see Appendix A.2) is in the range of several hundred milliseconds to several seconds, while a sound source signal may last from several minutes to several tens of minutes. If these convolutions are performed using Eq. (10.2), a large number of operations is required, which is far from practical.

For example, if the sampling frequency is 48 kHz, the sound source signal is a musical piece of 60 minutes in length, and the length of the impulse response is approximately 2.6 s (2^{17} samples), the required number of multiplications and additions are as follows:

$$M \times N = 48000 \times 60 \times 60 \times 2^{17} \cong 2.26 \times 10^{13} \quad (10.3)$$

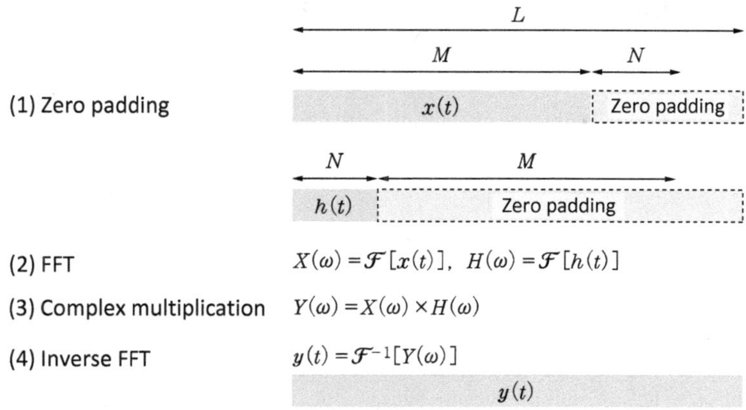

Fig. 10.10 Conceptual diagram of convolution in time domain

Convolution in the time domain is the multiplication of the complex spectrum of the source signal and the impulse response in the frequency domain, as follows:

$$Y(\omega) = X(\omega) \times H(\omega) \tag{10.4}$$

where $X(\omega) = \mathcal{F}(x(t)), \quad H(\omega) = \mathcal{F}(h(t))$.

Assuming L is a power of two, and L > M + N, the convolution can be performed using the following procedure in the frequency domain (Fig. 10.10).

(1) Perform zero padding for the tail of the response of x(t) and h(t) in order to make the length of each response L.
(2) Obtain X(ω) and H(ω) by FFT for each x(t) and h(t).
(3) Obtain Y(ω), the data length of which is L, by complex multiplication, as shown in Eq. (10.4).
(4) Obtain y(t), the data length of which is L, by an inverse FFT of Y(ω).

Here, we calculate the number of operations using the previous example. The number of FFT and inverse FFT calculations is $\frac{L}{2}\log_2 L$ for multiplication and $L\log_2 L$ for addition. The numbers of complex multiplication computations is four for multiplication and two for addition.

Since M + N = 48,000 × 60 × 60 + 2^{17} and L, the smallest power of two that exceeds this value, is 2^{28}, the number of multiplications required for the convolution is expressed by Eq. (10.5) and the number of additions is expressed by Eq. (10.6).

$$\begin{aligned} &\left(\frac{L}{2}\log_2 L\right) + \left(\frac{L}{2}\log_2 L\right) + 4L + \left(\frac{L}{2}\log_2 L\right) \\ &= L\left\{\frac{3}{2}(\log_2 L) + 4\right\} = 2^{28} \times 46 \cong 1.23 \times 10^{10} \end{aligned} \tag{10.5}$$

$$\begin{aligned} &L\log_2 L + L\log_2 L + 2L + L\log_2 L \\ &\quad = L\{3(\log_2 L) + 2\} = 2^{28} \times 86 \cong 2.31 \times 10^{10} \end{aligned} \tag{10.6}$$

This number of calculations is significantly fewer than that in the time domain (both multiplication and addition required 2.26×10^{13} calculations).

B. *Overlap-add method*

However, for the case in which the sound source signal x(t) is long, $(L - M)$ pieces of zero padding in the FFT of the impulse response shown in Fig. 10.8 are much greater than N, and therefore the processing is inefficient. An approach to divide the sound source signal by the length, N, of the impulse response h(t) and perform FFT (Fig. 10.11) was considered. This is referred to as the overlap-add method.

Assuming that N is a power of two, convolution can be performed by the following process.

(1) Divide the source signal by N (the length of the impulse response h(t)).
(2) N zeros are added to $x_1(t)$ and $h_1(t)$, the length of which is 2 N.
(3) Perform a FFT, complex multiplication, and inverse FFT in order to obtain the convolution result for the data length, 2 N.
(4) Steps (2) and (3) are performed for all sections of the source signals, and the results are then added by shifting by N samples

Here, we calculate the number of operations using the previous example. Since $L = 2\,N = 2^{18}$, the number of multiplications required for the convolution of the first

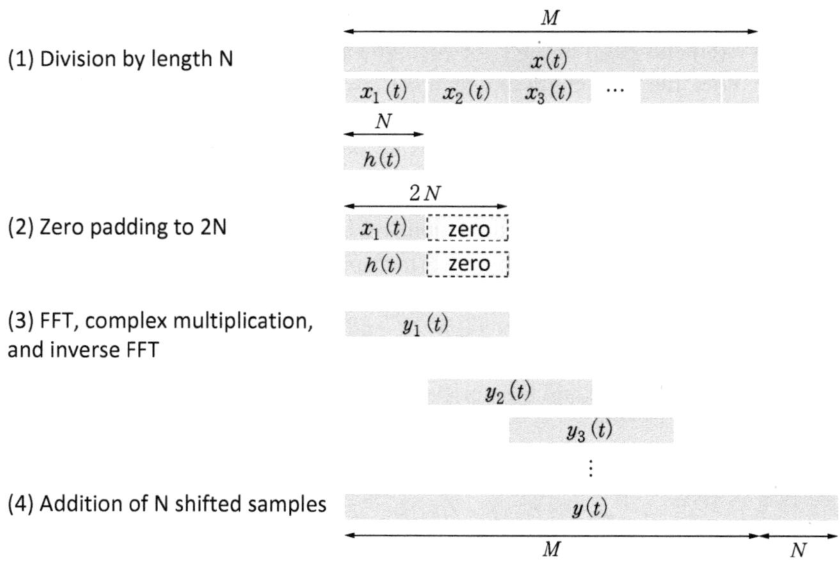

Fig. 10.11 Conceptual diagram of overlap-add method

section, $x_1(t)$ and $h_1(t)$, is expressed by Eq. (10.7), and the number of additions is expressed by Eq. (10.8).

$$L\left\{\frac{3}{2}(\log_2 L) + 4\right\} = 2^{18} \times 31 \cong 8.13 \times 10^6 \tag{10.7}$$

$$L\{3(\log_2 L) + 2\} = 2^{18} \times 56 \cong 1.47 \times 10^7 \tag{10.8}$$

For the second and subsequent sections, the FFT of the impulse response is unnecessary, and therefore the number of multiplications required for each section is expressed as follows:

$$L\{(\log_2 L) + 4\} = 2^{18} \times 22 \cong 5.77 \times 10^6 \tag{10.9}$$

The number of additions, including additions shifted by N samples, is expressed as:

$$L\{2(\log_2 L) + 2\} + 2^{17} = 2^{18} \times (2 \times 18 + 2) + 2^{17} \cong 1.01 \times 10^7 \tag{10.10}$$

Therefore, if all of these multiplications and additions are possible in the time period for N samples (2.6 s in this example), in other words, if 45 multiplications and 78 additions are possible for one sampling period (1/48000 s in this example), then the convolution can be performed in real time.

Moreover, since the number of sections after the second section is expressed as:

$$\frac{48000 \times 60 \times 60}{2^{17}} - 1 \cong 1317 \tag{10.11}$$

the number of multiplication and addition operations required for processing all sections is expressed as:

$$8.13 \times 10^6 + 5.77 \times 10^6 \times 1317 \cong 7.61 \times 10^9 \tag{10.12}$$

$$1.47 \times 10^7 + 1.01 \times 10^7 \times 1317 \cong 1.33 \times 10^{10} \tag{10.13}$$

In this example, the ratio of the number of computations of the overlap-add method to the conventional FFT is $7.61 \times 10^9/1.23 \times 10^{10} \cong 0.62$ for multiplication and $1.33 \times 10^{10}/2.31 \times 10^{10} \cong 0.58$ for addition.

References

Gardner B, Gardner S (1973) Problem of localization in the median plane: effect of pinna cavity occlusion. J Acoust Soc Am 53:400–408

Iida K, Oota M (2018) Median plane sound localization using early head-related impulse response. Appl Acoust 139:14–23
Iida K, Yairi M, Morimoto M (1998) Role of pinna cavities in median plane localization. Proceedings 16th Int Congress Acoustics:845–846
Iida K, Ishii Y, Nishioka S (2014) Personalization of head—related transfer functionsin the median plane based on the anthropometry of the listener's pinnae. J Acoust Soc Am 136:317–333
Musicant A, Butler R (1984) The influence of pinnae-based spectral cues on sound localization. J Acoust Soc Am 75:1195–1200
Shaw EAG, Teranishi R (1968) Sound pressure generated in an external-ear replica and real human ears by a nearby point source. J Acoust Soc Am 4:240–249
Takemoto H, Mokhtari P, Kato H, Nishimura R, Iida K (2012) Mechanism for generating peaks and notches of head-related transfer functions in the median plane. J Acoust Soc Am 132:3832–3841

Chapter 11
Comparison of HRTF Databases

Abstract Several research institutes have released databases of HRTFs. This chapter introduces representative databases and compares them from the viewpoints of spectral cues and pinna shape.

11.1 Representative HRTF Database

Table 11.1 shows representative publically available sites of HRTF databases.

In this chapter, the databases of the following five research institutes are compared.

1). Acoustics Research Institute (ARI), Austria
2). Center for Image Processing and Integrated Computing Interface Laboratory (CIPIC), U.S.A.
3). Spatial Hearing Laboratory (SHL), Chiba Institute of Technology, Japan
4). Institut de Recherche et Coordination Acoustique/Musique (IRCAM), France
5). Research Institute of Electrical Communication (RIEC), Tohoku University

An outline of these databases is shown in Table 11.2 (Yan et al. 2014). All research institutes measured HRTFs under the blocked-entrance condition (Shaw and Teranishi 1968).

The minimum number of subjects is 45 (CIPIC) and the maximum number of subjects is 105 (RIEC). In the four research institutes other than IRCAM, an array in which multiple loudspeakers are arranged in the vertical direction is installed, and the HRIRs for various three-dimensional directions were measured by rotating the subject or the array in the horizontal direction. At IRCAM, HRIRs were measured by moving one loudspeaker in the vertical direction and then rotating the subject in the horizontal direction.

There are large differences in the lengths of the measured HRIRs, which are 200 samples in CIPIC and 8192 samples in IRCAM. The number of measurement directions is seven to 148 at SHL, 1250 at CIPIC, 1550 at ARI, 187 at IRCAM, and 865 at RIEC.

K. Iida, *Head-Related Transfer Function and Acoustic Virtual Reality*,
https://doi.org/10.1007/978-981-13-9745-5_11

Table 11.1 Publically available sites of HRTF databases

Institute	Country	URL
ARI	Austria	https://www.kfs.oeaw.ac.at/index.php?lang=en
IRCAM	France	http://recherche.ircam.fr/equipes/salles/listen/
CIPIC	USA	http://interface.cipic.ucdavis.edu/sound/hrtf.html
MIT	USA	http://sound.media.mit.edu/resources/KEMAR.html
SHL	Japan	http://www.iida-lab.it-chiba.ac.jp/HRTF/
RIEC	Japan	http://www.ais.riec.tohoku.ac.jp/lab/db-hrtf/index-j.html
Nagoya	Japan	http://www.sp.m.is.nagoya-u.ac.jp/HRTF/index-j.html
ITA	Germany	http://gershwin.akustik.rwth-aachen.de/hrtf/hrtf-lic.php

Table 11.2 Outline of the five HRTF databases considered herein. (Yan et al. 2014)

	ARI	CIPIC	SHL	IRCAM	RIEC
Subject number	82	45	61	50	105
Source signal	ML sequence	MESM	swept–sine	OATSP	log sweep
Data length	256	200	512	8192	512
Sampling frequency (Hz)	48,000	44,100	48,000	44,100	48,000
Number of directions	1550	1250	7–148	187	865
Microphone	KE–4–211–2	ER–7C	WM64AT102	FG3329	FG3329
Manufacturer	Sennhiser	Etymtic	Panasonic	Knowles	Knowles
Loudspeaker	10 BGS	Acoustimass™	FE83E	system600	FE83E
Manufacturer	VIFA	Bose	Fostex	TANNOY	Fostex
Number of pieces	22	5	7	1	35
Data format	mat	mat	bin	mat	SOFA

11.2 Comparison of Spectral Cues

Next, N1, N2, and P1 frequency are compared between databases. The N1, N2 and P1 frequencies were calculated from the HRIRs for the front direction in the five databases using the method described in Sect. 10.2. Their histograms are shown in Fig. 11.1 (Yan et al. 2014).

The histograms for each database are approximately normally distributed. However, the peaks of the ARI histograms are at a higher frequency than the other histograms.

The average, minimum, and maximum frequencies of N1, N2 and P1 for the front direction of each database are shown in Table 11.3. Comparing the databases, the average frequency of RIEC is the lowest, and that of ARI is the highest for N1 and P1. For N2, the average frequency of SHL is the lowest, and that of ARI is the highest.

As such, the N1, N2, and P1 frequencies in Japanese databases are low, and those in ARI are high, as compared to the other databases.

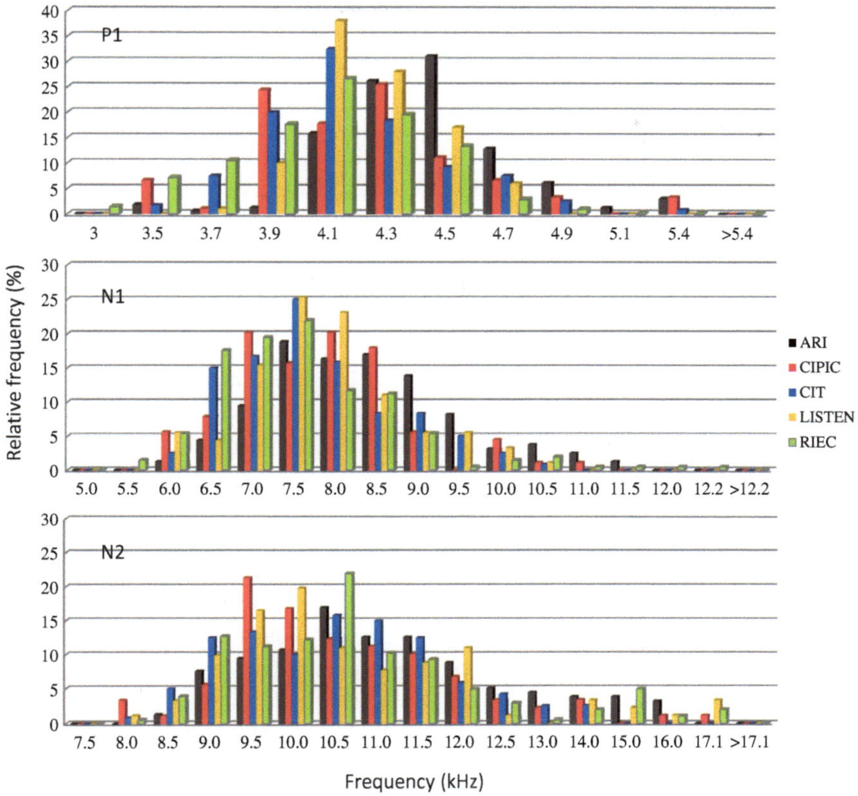

Fig. 11.1 Histograms of P1, N1, and N2 frequencies of HRTFs for front direction for five HRTF databases. (Yan et al. 2014)

Table 11.3 Average, minimum, and maximum frequencies of N1, N2, and P1 for the front direction for five database (Hz). (Yan et al. 2014)

	ARI	CIPIC	SHL	IRCAM	RIEC
P1 Ave	4333	4095	4059	4131	3969
P1 Min	3281	3187	3469	3618	2438
P1 Max	5250	5340	5250	4651	4875
N1 Ave	8101	7545	7481	7585	7301
N1 Min	6000	5771	5531	5685	5063
N1 Max	11,250	10,939	10,031	10,164	12,188
N2 Ave	10,959	10,384	10,287	10,519	10,549
N2 Min	8063	7494	7781	7752	7688
N2 Max	15,938	16,107	13,500	16,882	17,063

Table 11.4 Results of statistical tests for N1, N2, and P1 frequency. (Yan et al. 2014)

N1	ARI	CIPIC	SHL	IRCAM	RIEC
CIPIC	**	–			
CIT	**		–		
LISTEN	**			–	
RIEC	**			*	–
N2	ARI	CIPIC	SHL	IRCAM	RIEC
CIPIC	**	–			
CIT	**		–		
LISTEN	**			–	
RIEC	**				–
P1	ARI	CIPIC	SHL	IRCAM	RIEC
CIPIC	**	–			
CIT	**		–		
LISTEN	**		*	–	
RIEC	**	**	*	**	–

** $p < 0.01$; * $p < 0.05$

Furthermore, statistical tests were performed in order to verify whether there exists a significant difference in the average frequencies among the databases. The results are shown in Table 11.4.

The average frequency of ARI was significantly higher for N1, N2, and P1 compared to the other four databases ($p < 0.01$). For P1, the average frequency of RIEC is significantly lower compared with the other four databases ($p < 0.05$).

11.3 Comparison of Pinna Shape

As mentioned in Chap. 3, since N1, N2, and P1 are caused by the resonance of the cavities in the pinna, the differences in N1, N2, and P1 frequencies among the databases are related to the differences in pinna size.

Among the five HRTF databases, for which the N1, N2, and P1 frequencies were analyzed, detailed pinna anthropometric dimension data for the subjects are available for ARI, CIPIC, and SHL. The number of subjects for ARI, CIPIC, and SHL are 40 (80 ears), 34 (74 ears), and 28 (56 ears), respectively.

Histograms of each pinna anthropometric dimension are shown in Fig. 11.2, and their statistics are shown in Table 11.5. The average pinna anthropometric dimension of SHL is larger than those of ARI and CIPIC, except for x7.

Statistical tests were performed in order to verify whether there exists a significant difference in the average pinna anthropometric dimensions among the databases. The results are shown in Table 11.6. Significant differences ($p < 0.05$) were observed

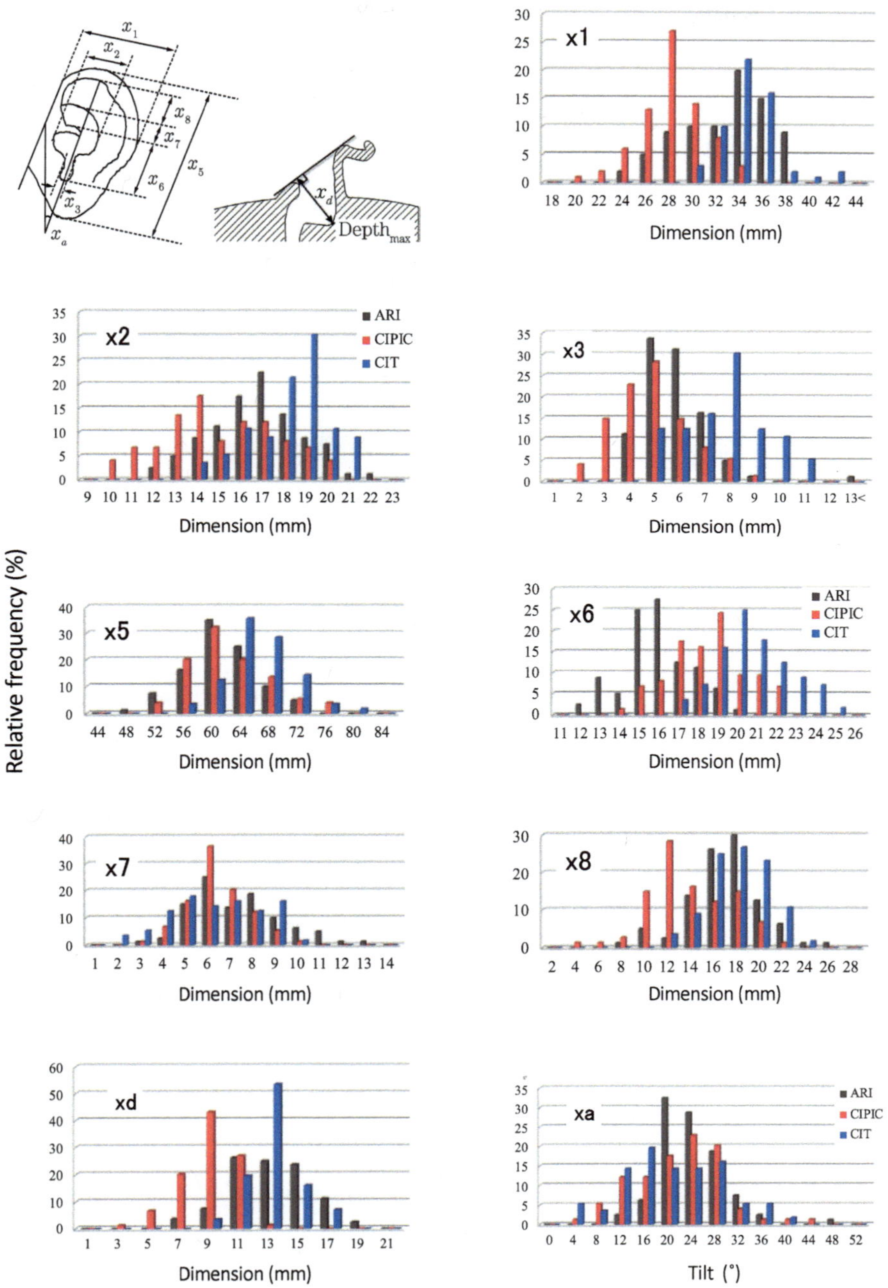

Fig. 11.2 Histograms of pinna anthropometric dimensions. (Yan et al. 2014)

Table 11.5 Statistics for pinna anthropometric dimensions (mm). (Yan et al. 2014)

		x_1	x_2	x_3	x_5	x_6	x_7	x_8	x_d	x_a[°]
Mean	ARI	33.48	16.90	6.20	62.43	16.22	7.51	17.72	13.71	25.48
	CIPIC	29.05	15.45	5.40	63.96	18.87	6.83	14.80	9.76	23.29
	SHL	35.48	18.68	8.19	68.26	21.15	6.69	18.92	13.93	21.86
Min	ARI	25.00	12.10	4.00	48.00	12.10	3.80	9.00	7.00	12.00
	CIPIC	21.84	10.37	2.70	54.24	14.32	3.78	5.81	3.65	5.94
	SHL	31.22	14.78	5.34	58.23	17.72	2.58	13.24	9.71	4.00
Max	ARI	39.20	22.00	18.00	74.00	20.00	13.20	26.70	19.20	49.00
	CIPIC	35.34	20.97	9.11	79.55	22.94	10.46	22.37	13.11	44.48
	SHL	43.83	21.84	11.88	83.18	25.09	10.28	24.14	17.60	40.00
Standard deviation	ARI	3.60	2.06	1.67	5.00	1.70	1.99	3.07	2.67	5.57
	CIPIC	2.74	2.58	1.51	5.58	1.96	1.32	3.51	1.79	7.60
	SHL	2.36	1.71	1.65	4.66	1.80	1.99	2.50	1.74	8.37

Table 11.6 Results of statistical tests for pinna anthropometric dimensions. *: $p < 0.05$, **: $p < 0.01$ (Yan et al. 2014)

Pinna anthropometry	Comparison between ARI and CIPIC	ARI and SHL	CIPIC and SHL	Size relationship				
x_1	**	**	**	CIPIC	<	ARI	<	SHL
x_2	**	**	**	CIPIC	<	ARI	<	SHL
x_3	**	**	**	CIPIC	<	ARI	<	SHL
x_5		**	**	ARI	≅	CIPIC	<	SHL
x_6	**	**	**	ARI	<	CIPIC	<	SHL
x_7	*	*		SHL	≅	CIPIC	<	ARI
x_8	**	*	**	CIPIC	<	ARI	<	SHL
x_d	**		**	CIPIC	<	ARI	≅	SHL
x_a	*	**		SHL	≅	CIPIC	<	ARI

for almost all combinations. In other words, there exists a difference in ear size among the databases.

In the previous section, we found that N1, N2, and P1 frequencies were higher for ARI than for the other four databases. Here, let us consider the reasons. Since N1, N2, and P1 are generated by resonance in the pinna cavities, it is inferred that the pinna dimensions of ARI are smaller than those in other databases. The pinna anthropometric parameter for which the average dimension of ARI was statistically significantly smaller than in other databases was x6 (length of cavity of concha) ($p < 0.01$). As described in Sect. 3.5, the x6 dimension is shown to have a significant effect on the N1, N2, and P1 frequencies. In other words, the fact that x6 is smaller than other databases is considered to be one of the reasons why the N1, N2, and P1 frequencies for ARI are higher than for other databases.

References

Shaw EAG, Teranishi R (1968) Sound pressure generated in an external-ear replica and real human ears by a nearby point source. J Acoust Soc Am 4:240–249

Yan X, Iida K, Ishii Y (2014) Comparison in frequencies of spectral peaks and notches and anthropometric of pinnae between HRTH databases. IEICE Technical Report, EA2014–19. pp. 43–48

Chapter 12
Principle of Three-Dimensional Sound Reproduction

Abstract By appropriately using the knowledge on HRTFs described in detail above, the three-dimensional acoustical sensation of an existing sound field can be reproduced or that of an imaginary sound field can be generated beyond time and space. This chapter introduces the principle of three-dimensional sound reproduction through headphones and with two loudspeakers.

12.1 Reproduction of Ear Input Signals through Headphones

12.1.1 Basic Principle

This section focuses on the case reproducing the ear input signal, X(ω), recorded by the dummy head in the original sound field through headphones, as shown in Fig. 12.1.

If the signal recorded by the dummy head is presented to the listener as it is, the transfer function, H(ω), from the headphones to the entrance of the ear canal of the listener affects the listener's ear input signal, Y(ω). This transfer function does not exist when recording with a dummy head in the original sound field. In order to reproduce the ear-input signals in the original sound field at the entrance of the ear canal of the listener in the playback sound field, this transfer function must be removed.

Therefore, the ear-input signal, X(ω), which is recorded by the dummy head in the original sound field, is processed by the filter, G(ω), and then reproduced at the entrance of the ear canal of the listener through headphones, as shown in Fig. 12.2.

The reproduced ear-input signal Y(ω) is expressed as follows:

$$Y(\omega) = X(\omega) \cdot G(\omega) \cdot H(\omega) \tag{12.1}$$

Thus, the filter G (ω), which equalizes Y(ω) to X (ω), is rewritten as follows:

K. Iida, *Head-Related Transfer Function and Acoustic Virtual Reality*,
https://doi.org/10.1007/978-981-13-9745-5_12

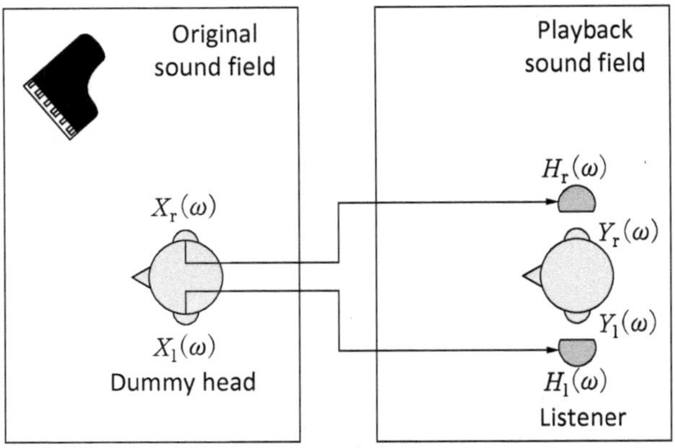

Fig. 12.1 Reproduction of ear input signal recorded by dummy head in original sound field through headphones

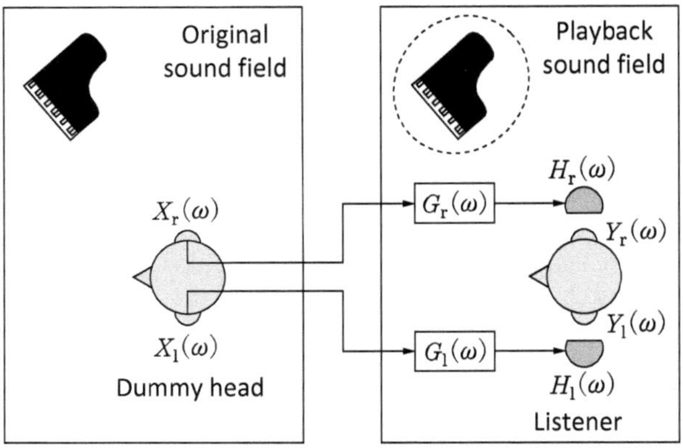

Fig. 12.2 Reproduction of ear input signal recorded by dummy head in original sound field through headphones with compensation filter G

$$X(\omega)\angle\varphi(\omega) = X(\omega) \cdot G(\omega) \cdot H(\omega) \tag{12.2}$$

Then, the following equation is obtained:

$$G(\omega) = \frac{\angle\varphi(\omega)}{H(\omega)} \tag{12.3}$$

where $\angle\varphi$ (ω) is the linear phase delay introduced so that the filter, G(ω), satisfies the law of causality.

Applying the same phase delay, $\angle\varphi_0(\omega)$, to the left and right ears, the signal Y(ω) at the entrance of the ear canal of the listener can be equalized to the signal X(ω) recorded by the dummy head, and interaural difference information for signal Y(ω) can also be equalized to that for signal X(ω), as shown by Eq. (12.4)

$$\frac{Y_r(\omega)}{Y_l(\omega)} = \frac{X_r(\omega)\angle\varphi_r(\omega)}{X_l(\omega)\angle\varphi_l(\omega)} = \frac{X_r(\omega)\angle\varphi_0(\omega)}{X_l(\omega)\angle\varphi_0(\omega)} = \frac{X_r(\omega)}{X_l(\omega)} \tag{12.4}$$

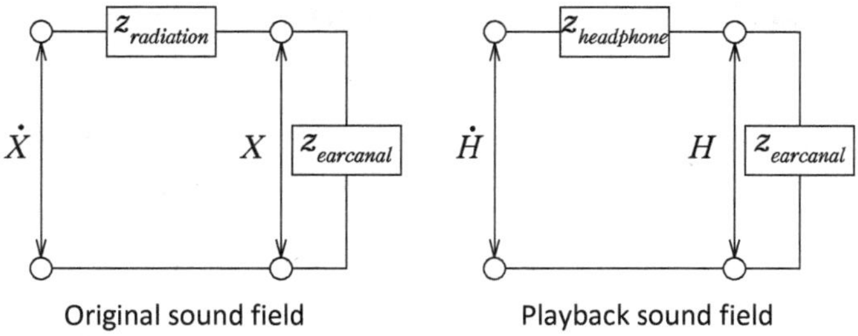

Fig. 12.3 Relationship between ear-input signal with closed entrance of ear canal and that with open entrance of ear canal

Processing the ear input signal X(ω) recorded with the dummy head, which is strikingly similar to the listener's head, with the filter G(ω) and reproducing the processed ear input signal through headphones, the acoustic signal to be heard in the original sound field would be provided to the listener.

Then, assuming the acoustic impedance of the ear canal of such a dummy head to be $Z_{earcanal}(\omega)$ and the radiation impedance as viewed from the sound source side from the entrance of the ear canal to be $Z_{radiation}(\omega)$, the relationship between the ear input signal $\dot{X}(\omega)$, for which the influence of sound propagation characteristics in the ear canal is removed by closing the entrance of the canal of the dummy head, and the ear input signal X(ω) recorded at the opened entrance of the ear canal is expressed as follows (Eq. (12.5a) and Fig. 12.3(a)).

$$\dot{X}(\omega) = \frac{Z_{earcanal}(\omega) + Z_{radiation}(\omega)}{Z_{earcanal}(\omega)} X(\omega) \tag{12.5a}$$

In the same way, assuming the acoustic impedance of the ear canal of a listener to be $\dot{Z}_{earcanal}(\omega)$ ($= Z_{earcanal}(\omega)$) and the radiation impedance as viewed from the sound source side from the entrance of the ear canal to be $Z_{headphone}(\omega)$, the relationship between the transfer function from the headphones to the closed entrance of the ear canal $\dot{H}(\omega)$ and the transfer function from the headphones to the opened entrance of the ear canal $H(\omega)$ is given in Eq. (12.5b) and in Fig. 12.3(b).

$$\begin{aligned} \dot{H}(\omega) &= \frac{\dot{Z}_{earcanal}(\omega) + Z_{headphone}(\omega)}{\dot{Z}_{earcanal}(\omega)} H(\omega) \\ &= \frac{Z_{earcanal}(\omega) + Z_{headphone}(\omega)}{Z_{earcanal}(\omega)} H(\omega) \end{aligned} \tag{12.5b}$$

Moreover, applying Eq. (12.5a) to Eq. (12.1), the filter $\dot{G}(\omega)$, which reproduces the ear input signal X(ω) with the open ear canal or X(ω)∠φ(ω), is obtained from the ear input signal $\dot{X}(\omega)$ recorded with the closed ear canal of the dummy head.

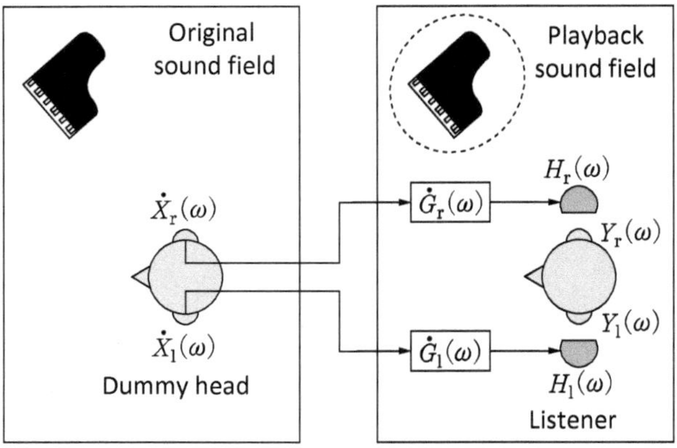

Fig. 12.4 Reproduction of ear input signal recorded at closed entrance of ear canal of dummy head through headphones with compensation filter $\dot{G}(\omega)$

$$\begin{aligned}\dot{G}(\omega) &= \frac{\angle\varphi(\omega)}{H(\omega)}\cdot\frac{X(\omega)}{\dot{X}(\omega)}\\ &= G(\omega)\cdot\frac{Z_{earcanal}(\omega)}{Z_{earcanal}(\omega)+Z_{radiation}(\omega)}\\ &= G(\omega)\cdot\frac{\dot{Z}_{earcanal}(\omega)}{\dot{Z}_{earcanal}(\omega)+Z_{radiation}(\omega)}\end{aligned} \tag{12.6}$$

Processing the ear input signal X(ω) recorded at the closed entrance of the ear canal of the dummy head, which is strikingly similar to the listener's head, with the filter $\dot{G}(\omega)$ and reproducing the processed ear input signal through headphones, the acoustic signal to be heard in the original sound field would be provided to the listener, as shown in Fig. 12.4.

Furthermore, applying Eq. (12.5b) to Eq. (12.6), the filter $\dot{G}(\omega)$ is expressed as follows:

$$\begin{aligned}\dot{G}(\omega) &= \frac{\angle\varphi_0(\omega)}{\dot{H}(\omega)}\cdot\frac{\dot{Z}_{earcanal}(\omega)+Z_{headphone}(\omega)}{\dot{Z}_{earcanal}(\omega)+Z_{radiation}(\omega)}\\ &\triangleq \frac{\angle\varphi_0(\omega)}{\dot{H}(\omega)}\cdot PDR(\omega)\end{aligned} \tag{12.7}$$

Using free air equivalent coupling to the ear (FEC) headphones, for which the pressure distribution ratio (PDR) of the second term of the right-hand side of Eq. (12.7) is regarded as approximately unity (Møller 1992), the ear input signal $\dot{X}(\omega)$, and the inverse filter $\angle\varphi_0(\omega)/\dot{H}(\omega)$ of a transfer function between the headphones and the listener's closed entrance of the ear canal, the listener is given the same acoustic signal as that heard in the original sound field.

The PDR can also be expressed as the ratio of sound pressure as in the following equation:

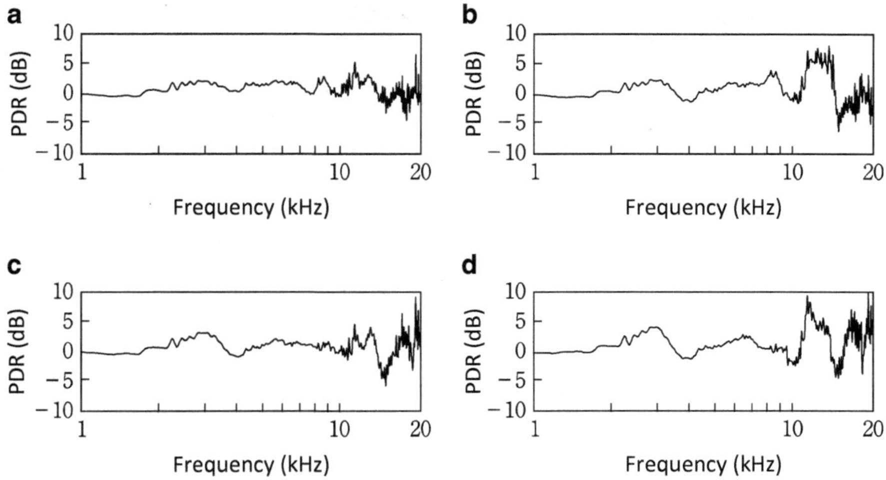

Fig. 12.5 Measured PDRs for four kinds of open-air headphones. (**a**) K1000 (AKG), (**b**) DT990 PRO (beyerdynamic), (**c**) ATH-AD700 (audio-technica), and (**d**) CD900 (Sony)

$$PDR(\omega) = \frac{\dot{Z}_{earcanal}(\omega) + Z_{headphone}(\omega)}{\dot{Z}_{earcanal}(\omega) + Z_{radiation}(\omega)} = \frac{P3(\omega)/P2(\omega)}{P6(\omega)/P5(\omega)} \tag{12.8}$$

where P2 and P3 are the sound pressures measured at the closed and open entrances of the ear canal for a sound source presented from a loudspeaker, and P5 and P6 are the sound pressures measured at the closed and open entrances of the ear canal for a sound source presented from headphones.

Figure 12.5 shows the measured PDR for four kinds of open-air headphones (K1000 (AKG), DT990 PRO (beyerdynamic), ATH-AD700 (audio-technica), and CD900 (Sony)). The value of the PDR becomes large, and the fluctuation is also large at 10 kHz or more for all of these headphones. Below 10 kHz, the value is close to 0 dB for K1000 and DT990 PRO. The relative relationship of frequency average (RMS value) was as follows: K1000 < DT990 PRO < AD700 < CD900.

The compensation process of $\dot{H}(\omega)$ for two headphones, which are regarded as FEC headphones, i.e., K1000 and DT990 PRO, is described. For K1000, $\dot{H}(\omega)$ can be compensated as follows.

1. The earplug-type microphones are placed into the ear canals of the subject. Note that the diaphragms of the microphones are located at the entrances of the ear canals.
2. The subjects wear the open-air headphones, and maximum-length sequence signals (48-kHz sampling, 12th order, and no repetitions) are emitted through the headphones. The signals are received by the earplug-type microphones, and the transfer functions between the open-air headphones and the earplug-type microphones are obtained.

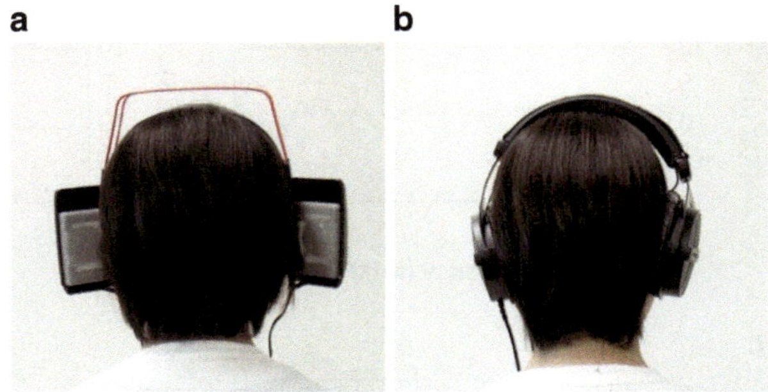

Fig. 12.6 Subject wearing headphones (**a**) K1000 (AKG) (**b**) DT990 PRO (beyerdynamic)

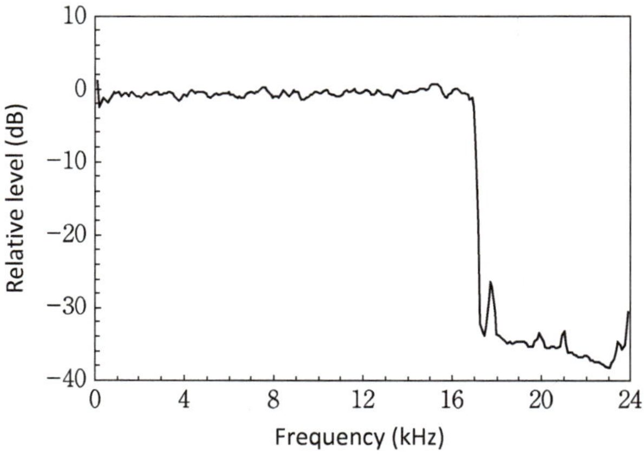

Fig. 12.7 Transfer function between headphones and earplug-type microphone with compensation filter, $\dot{G}(\omega)$ (Iida and Ishii 2018)

3. The earplug-type microphones are then removed without displacing the headphones because the pinnae of the subject are not enclosed by the headphones, as shown in Fig. 12.6(a).
4. The acoustic signals with the inverse filter $\angle\varphi_0(\omega)/\dot{H}(\omega)$ are delivered through the headphones.

The typical peak-to-peak range of the transfer functions between the headphones and the earplug-type microphones from 200 Hz to 17 kHz was approximately 20 dB. This was reduced to 3 dB by the compensation filter, $\dot{G}(\omega)$, as shown in Fig. 12.7.

For DT990 PRO, since the pinnae are enclosed by the headphones (Fig. 12.6(b)), putting on and taking off the headphones are necessary in order to measure $\dot{H}(\omega)$. Since $\dot{H}(\omega)$ varies with the position of the headphones, exact cancelation of $\dot{H}(\omega)$ cannot be achieved. Fig. 12.8 shows an example of the measured $\dot{H}(\omega)$ of DT990 PRO for four subjects. The peak-to-peak range exceeds 30 dB, as reported previously (Møller et al. 1995).

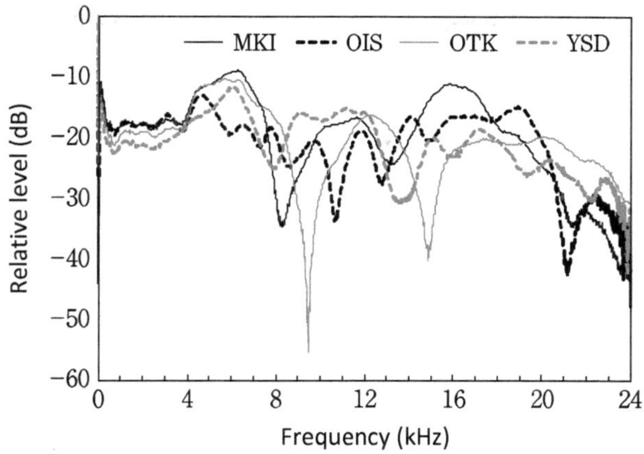

Fig. 12.8 Transfer function between DT990 headphones and earplug-type microphone for which compensation filter $\angle\varphi_0(\omega)/\dot{H}(\omega)$ is not included (four subjects)

12.1.2 Accuracy of Sound Image Localization

Here, sound localization accuracy in the median plane through headphones is introduced.

The results of the localization tests in the median plane through K1000 headphones with a compensation filter $\angle\varphi_0(\omega)/\dot{H}(\omega)$ and through DT990 PRO headphones without a compensation filter are shown in Figs. 12.9 and 12.10, respectively. The subject's own measured HRTFs were used.

For K1000 headphones with a compensation filter, most of the responses were distributed over a diagonal line (Fig. 12.9). However, subject OTK responded around 90° for target vertical angles of 60°, 120°, and 150°.

For DT990 PRO headphones without a compensation filter, most of the responses of subject MKI were distributed along a diagonal line (Fig. 12.10). However, for subject OTK, the responses were distributed widely from above to behind for the target vertical angles of 30° to 150°.

As described above, for headphones of a normal type such as DT 990 PRO, exact compensation is not possible because it is necessary to remove and attach the headphones in the compensation process. However, sound image localization tests have been conducted with a compensation filter based on the subject's own characteristics of $\angle\varphi_0(\omega)/\dot{H}(\omega)$ averaged over several measurements (Møller et al. 1996).

The results are shown in Fig. 12.11(a). Note that the HRTFs used are not those of the subjects themselves but rather are those of a typical subject introduced in Sect. 4.3.2. Although front-back error occurred in the median plane (front low), rising of a sound image did not occur.

Figure 12.11(b) shows the results of sound image localization tests, in which the compensation was performed using a filter based on the characteristics of $\angle\varphi_0(\omega)/\dot{H}(\omega)$ averaged among several subjects.

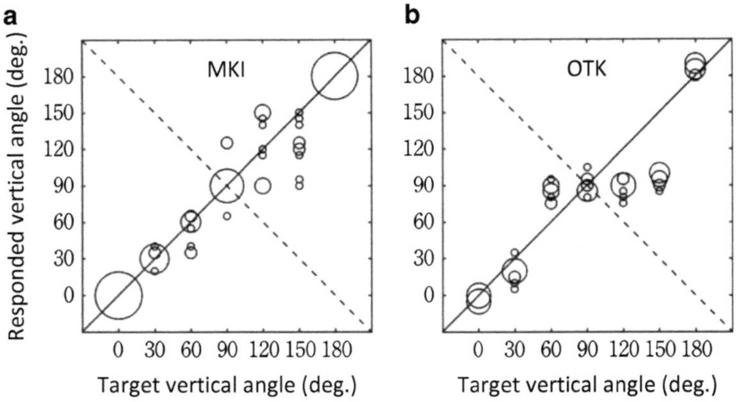

Fig. 12.9 Responses to subject's own measured HRTFs in upper median plane through K1000 headphones with compensation filter $\angle\varphi_0(\omega)/\dot{H}(\omega)$. (**a**) subject MKI and (**b**) subject OTK

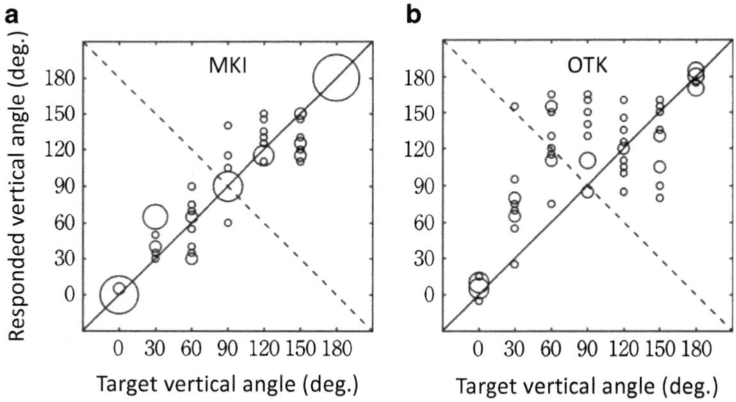

Fig. 12.10 Responses to subject's own measured HRTFs in upper median plane through DT990 PRO headphones without compensation filter. (**a**) subject MKI and (**b**) subject OTK

There seems to be no remarkable difference between (a) and (b). However, there exists a statistically significant difference in the front-back error ratio ($p < 0.05$), as shown in Table 12.1.

12.1.3 Introduction of Dynamic Cue

The effectiveness of changing the ear input signals following the head movement (dynamic cue) of the listener during ear input signal reproduction in front-back discrimination has been reported.

In the 1980s, a system that synchronizes the KEMAR dummy head with a subject's head movement was developed (Calhoun and Janson 1990), and similar

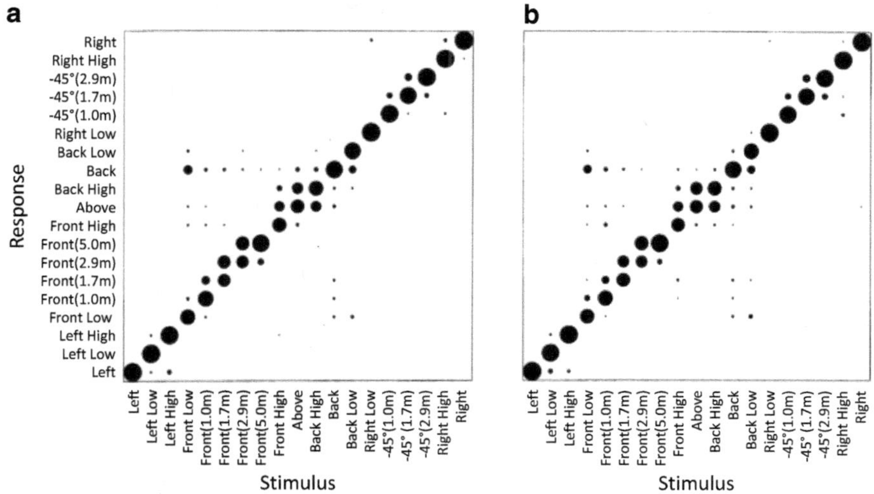

Fig. 12.11 Results of localization tests using compensation filter based on (**a**) subject's own characteristics of $\angle\varphi_0(\omega)/\dot{H}(\omega)$ averaged over several measurements and (**b**) on characteristics of $\angle\varphi_0(\omega)/\dot{H}(\omega)$ averaged among several subjects. (Møller et al. 1996). The values in parentheses indicate the distance of the sound source. The distance of the sound source without parenthesis was 1.0 m. Note that the HRTFs used are not those of the subjects themselves but rather those of a typical subject introduced in Sect. 4.3.2

Table 12.1 Front-back error ratio in median plane (%). (Møller et al. 1996)

Averaged over several measurements	Averaged among several subjects
21.2	24.0

dynamic dummy heads, the shape of the head and pinnae of which were reproduced for a specific listener, have been reported (Toshima et al. 2003).

However, these systems have a problem in that they are effective only while listening to the original sound field because synchronization of movement between the listener and the dummy head is important, and there is no effect on the signal once recorded (Fig. 12.12a).

If the temporal and spatial structures of the incident sounds in the original sound field are known, it is possible to capture the head movement of the listener with a head tracker and to change the HRTF following the head movement of the listener. However, since the incident sound structures of the original sound field are unknown in general, the dynamic cue is not applicable to three-dimensional acoustic reproduction of the original sound field. Its application is limited to the creation of a virtual sound field (Fig. 12.12b).

In order to solve such a problem, a system that provides ear-input signals that reflect HRTFs by synchronizing with a listener's head movement using a large number of microphones (e.g., 252 channels) placed on a sphere has been developed (Sakamoto et al. 2015).

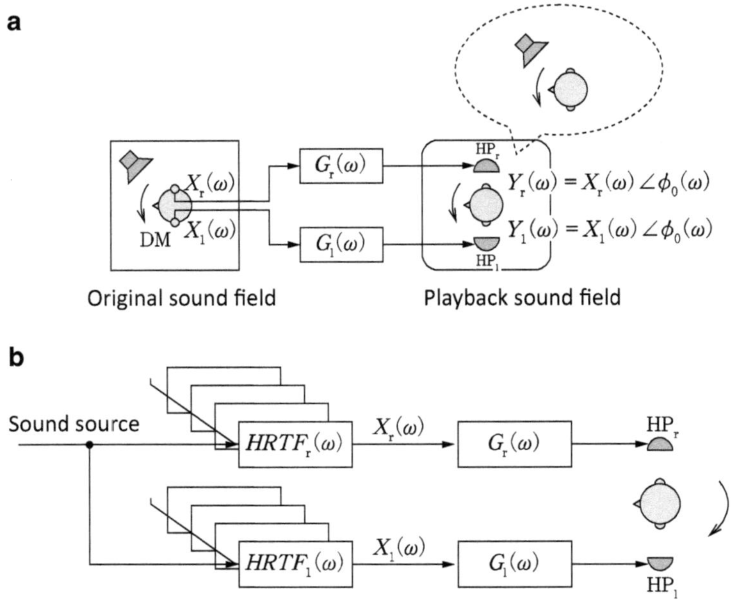

Fig. 12.12 Three-dimensional acoustic reproduction system following head movement of listener (**a**) Dummy head following subject's head movement (**b**) System that changes HRTF following head movement of listener

12.2 Reproduction of Ear-Input Signals with Two Loudspeakers

12.2.1 Basic Principle

A system that uses multiple loudspeakers and a digital filter matrix to reproduce the acoustic signals obtained at the entrances of the ear canals of a listener in the original sound field to the entrances of the ear canals of a listener in an arbitrary sound field is generally called a transaural system. Although a system using three or more loudspeakers has also been proposed, in the following, we outline a transaural system using two loudspeakers as a minimum configuration.

Schroeder and Atal (1963) were the first to propose a transaural system. They located two loudspeakers in a symmetrical position (±23°) in front of the listener and tried to reproduce the ear-input signals in a concert hall.

As shown in Fig. 12.13(a), assuming that the sound source in the original sound field is S, and the transfer function from the sound source to the listener's ears is H, the ear-input signal of a listener P is expressed as in Eq. (12.9). Here, H is a binaural room transfer function if the original sound field is a concert hall and an HRTF if the original sound field is an anechoic chamber that includes only direct sounds. The suffixes l and r indicate the left and right ears, respectively.

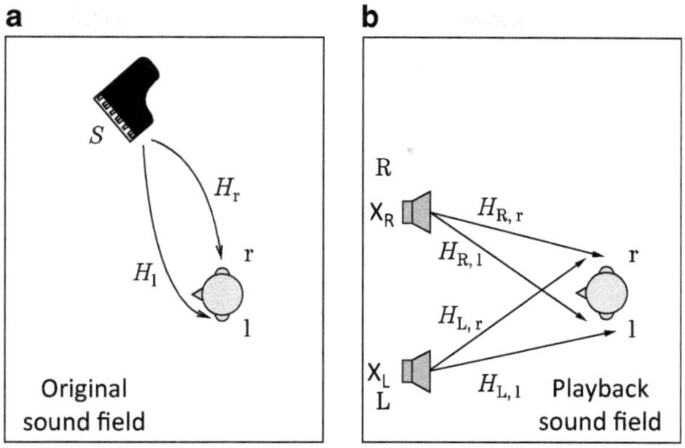

Fig. 12.13 Transfer functions in original sound field and playback sound field

On the other hand, when the signal X is emitted from the two loudspeakers in the playback sound field, as shown in Fig. 12.13(b), the ear-input signals of listener P′ are expressed by Eq. (12.10). Here, subscripts L and R indicate the left and right loudspeaker, respectively, and l and r indicate the left and right ears, respectively.

$$\begin{cases} P_l(\omega) = S(\omega) \cdot H_l(\omega) \\ P_r(\omega) = S(\omega) \cdot H_r(\omega) \end{cases} \tag{12.9}$$

$$\begin{cases} P'_l(\omega) = X_L(\omega) \cdot H_{L,l}(\omega) + X_R(\omega) \cdot H_{R,l}(\omega) \\ P'_r(\omega) = X_R(\omega) \cdot H_{R,r}(\omega) + X_L(\omega) \cdot H_{L,r}(\omega) \end{cases} \tag{12.10}$$

Assuming that P and P′ are equal, Eq. (12.11) holds.

$$\begin{cases} S(\omega) \cdot H_l(\omega) = X_L(\omega) \cdot H_{L,l}(\omega) + X_R(\omega) \cdot H_{R,l}(\omega) \\ S(\omega) \cdot H_r(\omega) = X_R(\omega) \cdot H_{R,r}(\omega) + X_L(\omega) \cdot H_{L,r}(\omega) \end{cases} \tag{12.11}$$

By solving Eq. (12.11) for the radiation signals of the left and right loudspeakers, X, Eq. (12.12) is obtained.

$$\begin{cases} X_L(\omega) = S(\omega) \times \dfrac{H_l(\omega)H_{R,r}(\omega) - H_r(\omega)H_{R,l}(\omega)}{H_{L,l}(\omega)H_{R,r}(\omega) - H_{L,r}(\omega)H_{R,l}(\omega)} \\ X_R(\omega) = S(\omega) \times \dfrac{H_r(\omega)H_{L,l}(\omega) - H_l(\omega)H_{L,r}(\omega)}{H_{L,l}(\omega)H_{R,r}(\omega) - H_{L,r}(\omega)H_{R,l}(\omega)} \end{cases} \tag{12.12}$$

By applying such signal processing to the sound source signal, in principle, the ear-input signals in the original sound field can be reproduced at the entrances of the ear canals of a listener in the playback sound field.

Furthermore, Eq. (12.12) is transformed as follows using Eq. (12.9):

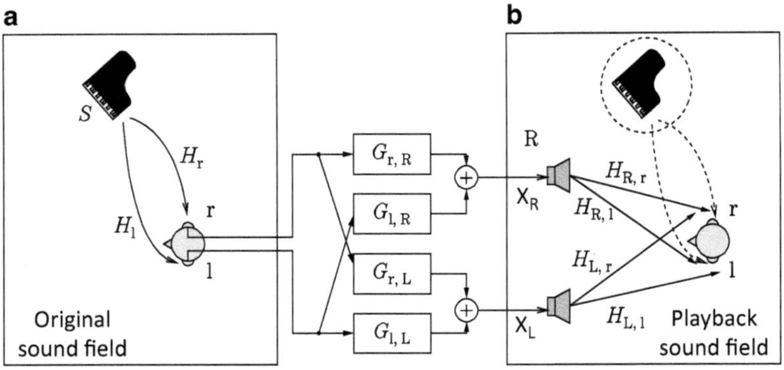

Fig. 12.14 Block diagram of transaural system. (**a**) original sound field and (**b**) playback sound field

$$\begin{cases} X_L(\omega) = \dfrac{P_l(\omega)H_{R,r}(\omega) - P_r(\omega)H_{R,l}(\omega)}{H_{L,l}(\omega)H_{R,r}(\omega) - H_{L,r}(\omega)H_{R,l}(\omega)} \\ X_R(\omega) = \dfrac{P_r(\omega)H_{L,l}(\omega) - P_l(\omega)H_{L,r}(\omega)}{H_{L,l}(\omega)H_{R,r}(\omega) - H_{L,r}(\omega)H_{R,l}(\omega)} \end{cases} \tag{12.13}$$

Equation (12.13) is equivalent to the filter matrix shown in Fig. 12.14. Each filter is expressed by Eq. (12.14).

$$\begin{cases} G_{l,L}(\omega) = \dfrac{H_{R,r}(\omega)}{H_{L,l}(\omega)H_{R,r}(\omega) - H_{L,r}(\omega)H_{R,l}(\omega)} \\ G_{r,L}(\omega) = \dfrac{-H_{R,l}(\omega)}{H_{L,l}(\omega)H_{R,r}(\omega) - H_{L,r}(\omega)H_{R,l}(\omega)} \\ G_{l,R}(\omega) = \dfrac{-H_{L,r}(\omega)}{H_{L,l}(\omega)H_{R,r}(\omega) - H_{L,r}(\omega)H_{R,l}(\omega)} \\ G_{r,R}(\omega) = \dfrac{H_{L,l}(\omega)}{H_{L,l}(\omega)H_{R,r}(\omega) - H_{L,r}(\omega)H_{R,l}(\omega)} \end{cases} \tag{12.14}$$

Although only the direct sound is drawn in Fig. 12.14 as a path from the sound source to the listener's ears in the original sound field, reflections may be included. In addition, reflections may be included in the playback sound field. However, if the impulse response from the loudspeaker to both ears in the playback sound field is long, the filter in Eq. (12.14) also requires a long response length. Generally, approximately four times the duration of the impulse response of the playback sound field is required for the filter.

12.2.2 *Accuracy of Sound Image Localization*

Thus, in principle, two loudspeakers associated with a digital filter matrix can reproduce the ear-input signals in the original sound field to both ears of a listener in any playback sound field.

However, in fact, in order to reproduce the three-dimensional sound field with two loudspeakers, it is necessary to satisfy the following two conditions; (1) use the listener's own HRTFs and (2) fix the position of the listener's head (ears). The first condition was mentioned in Chaps. 2 and 3, and therefore the findings on the robustness of the listening position are described here.

In the previously proposed transaural system, two loudspeakers were located at ±30° in front of the listener in the horizontal plane, but a method that places two loudspeakers in close proximity was proposed in consideration of the robustness of shifting the listening position (Kirkeby et al. 1998).

Moreover, a method of locating loudspeakers in a transverse plane, rather than a horizontal plane, has been investigated (Iida et al. 2012). Sound image localization tests were performed for a total of 17 loudspeaker placements (Fig. 12.15): 15 loudspeaker placements from ±20° through ±160° (T20 through T160) in 10° steps in the transverse plane, a conventional placement at ±30° in the horizontal plane (H30), and a close placement at ±6° in the horizontal plane (H6). The target directions were twelve azimuth angles in the horizontal plane (30° steps) and seven vertical angles in the upper median plane (30° steps).

The results of the sound image localization tests are shown in Figs. 12.16 through 12.19. For the target directions in the horizontal plane (Figs. 12.16 and 12.17), the loudspeaker placements for which the responses were distributed around the target directions were T60–T110. The loudspeaker placements for which there exist no statistically significant difference in mean localization error as compared with actual sound sources were T70, T80, and T100.

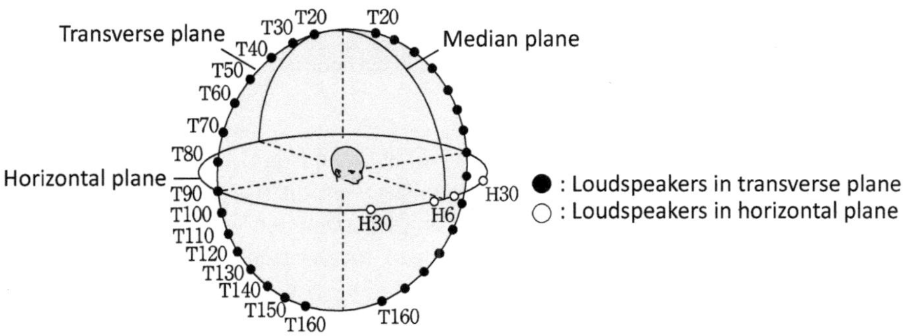

Fig. 12.15 Seventeen loudspeaker placements. (Iida et al. 2012)

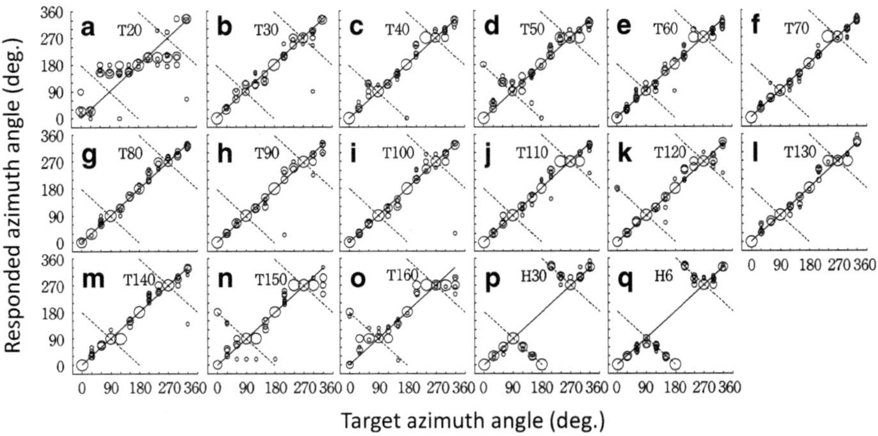

Fig. 12.16 Results of sound image localization tests in the horizontal plane, subject A. (Iida et al. 2012)

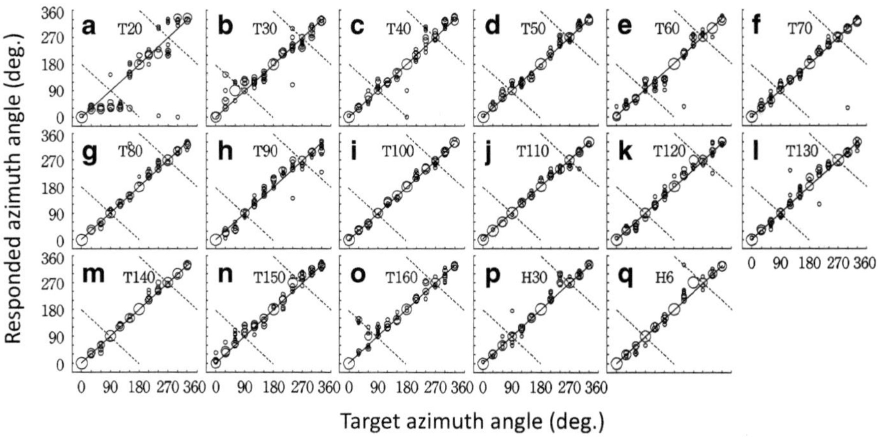

Fig. 12.17 Results of sound image localization tests in the horizontal plane, subject B. (Iida et al. 2012)

For the target directions in the upper median plane (Figs. 12.18 and 12.19), the loudspeaker placements for which the responses were distributed around the target directions were T60–T80. The loudspeaker placement for which there exists no statistically significant difference in the mean localization error compared with actual sound sources was T70.

These results infer that it is efficient to place loudspeakers of the transaural system diagonally upward in the transverse plane, rather than in the front direction in the horizontal plane, as originally proposed.

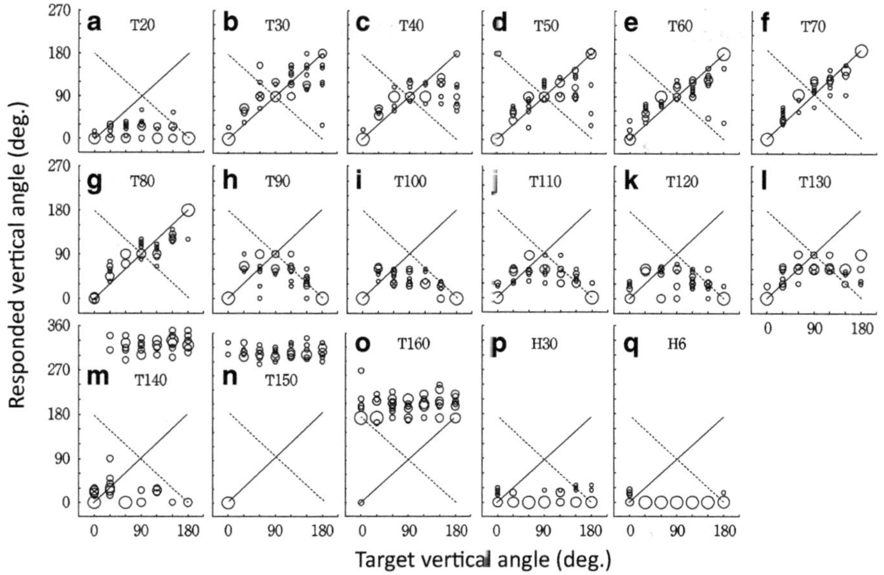

Fig. 12.18 Results of sound image localization tests in the upper median plane, subject A. (Iida et al. 2012)

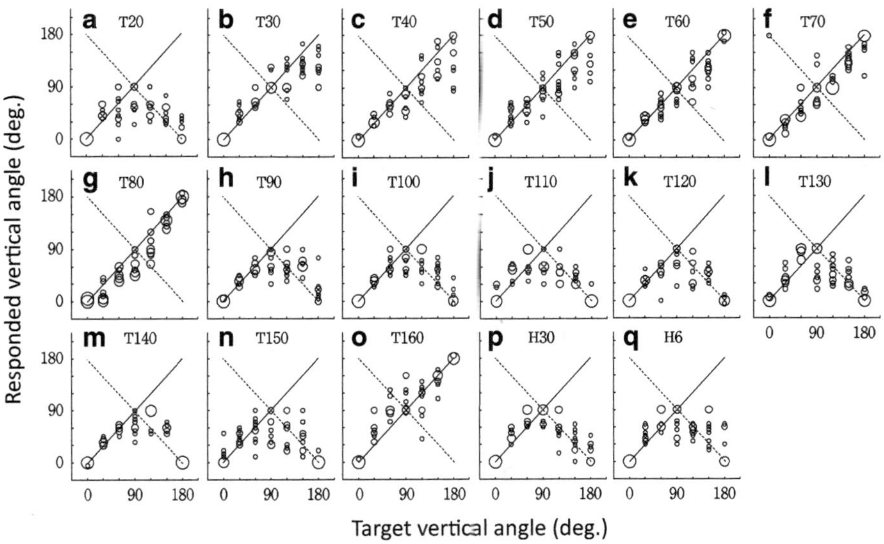

Fig. 12.19 Results of sound image localization tests in the upper median plane, subject B. (Iida et al. 2012)

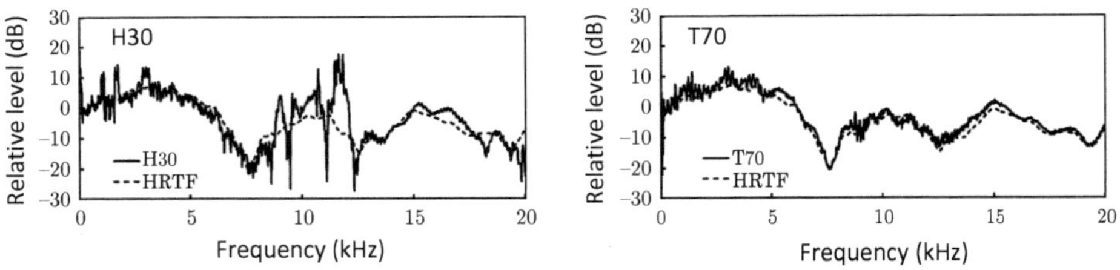

Fig. 12.20 Comparison of transfer functions. Solid line: reproduced, broken line: target. (Iida et al. 2012)

Figure 12.20 shows the target transfer functions and the reproduced transfer functions obtained by Eq. (12.11) for H30 and T70. The target direction is backward (180°). Remarkable notches and peaks that do not exist in the target transfer functions appeared at 8.5–11.0 kHz in the reproduced transfer function for H30. These notches and peaks are thought to be the reason why the subject did not perceive a sound image backward. On the other hand, a transfer function that was approximately the same as the target was reproduced for T70.

References

Calhoun GL, Janson WP (1990) Eye and head response as indicators of attention cue effectiveness. Proc. the human factors society 34th annual meeting. pp. 1–5

Iida K, Oota M (2018) Median plane sound localization using early head-related impulse response. Appl Acoust 139:14–23

Iida K, Ishii T, Ishii Y, Ikemi T (2012) Three-dimensional sound image control by two loudspeakers located in the transverse plane. J Acoust Soc Jpn 68:331–342. in Japanese

Kirkeby O, Nelson PA, Hamada H (1998) Local sound field reproduction using two closely spaced loudspeakers. J Acoust Soc Am 104:1973–1981

Møller H (1992) Fundamentals of binaural technology. Appl Acoust 36:171–218

Møller H, Hammershøi D, Jensen CB, Sørensen MF (1995) Transfer characteristics of headphones measured on human ears. J Audio Eng Soc 43:203–217

Møller H, Jensen CB, Hanmmershøi D, Sørensen MF (1996) Using a typical human subject for binaural recording. Audio Eng Soc Reprint 4157(C–10)

Sakamoto S, Hongo S, Okamoto T, Iwaya Y, Suzuki Y (2015) Sound–space recording and binaural presentation system based on a 252–channel microphone array. Acoust Sci Tech 36:516–526

Schroeder MR, Atal BS (1963) Computer simulation of sound transmission in rooms. Proc IEEE:536–537

Toshima I, Uematsu H, Hirahara T (2003) A steerable dummy head that tracks three–dimensional head movement: tele head. Acoust Sci Tech 24:327–329

Chapter 13
Acoustic VR System

Abstract An acoustic VR system provides a three-dimensional acoustical sensation of an existing sound field (as at concert halls, stadiums, or disaster sites) and/or an imaginary sound field (as in movies or video games), independent of time and space. This chapter introduces the system configuration and the applications of acoustic VR systems.

13.1 System Configuration

An example of a typical configuration of an acoustic VR system is shown in Fig. 13.1. The system consists of hardware (a PC, a digital audio interface, headphones, earplug-type microphones, and a head-tracker), software for signal processing, and a database (HRTF and pinna shape).

The main function of the acoustic VR system is to reproduce the ear-input signals obtained in an arbitrary sound field through headphones by the signal processing described in Chap. 12. This signal processing (convolution between a sound source signal and HRIRs) is performed on the PC.

Another function is to change the HRTFs to those of another direction in response to the head movement of a listener. In order to capture the direction of the listener's head, a head tracker is used. Rewriting of the HRTFs must be done within the threshold for the detection of system delay, i.e., 80 ms (Yairi et al. 2005).

Various systems have adopted an individualization function for HRTFs to ensure accurate sound image localization. An example system is shown in Fig. 13.2.

The external specifications of the acoustic VR system, i.e., the Sound Image Reproduction system with Individualized-HRTF, graphical User-interface and Successive head-movement tracking (SIRIUS), which was developed in the author's lab, are shown in Table 13.1.

This system runs on a Windows PC. An HRIR database (response length: 512 samples) is stored on the PC. A sound source signal and HRIRs are convolved in real time in order to control the direction and distance of a sound image.

K. Iida, *Head-Related Transfer Function and Acoustic Virtual Reality*,
https://doi.org/10.1007/978-981-13-9745-5_13

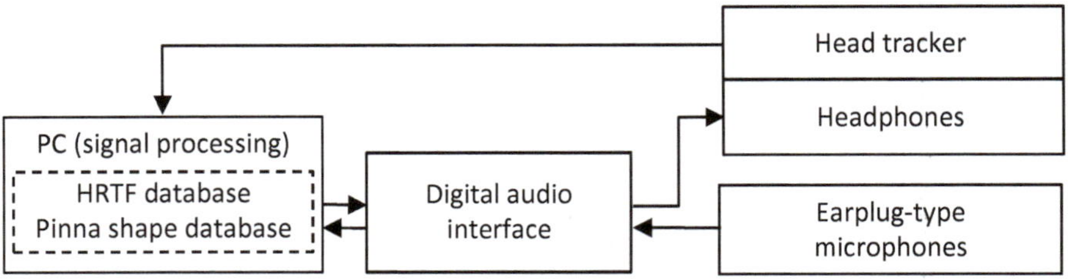

Fig. 13.1 Block diagram of three-dimensional auditory display

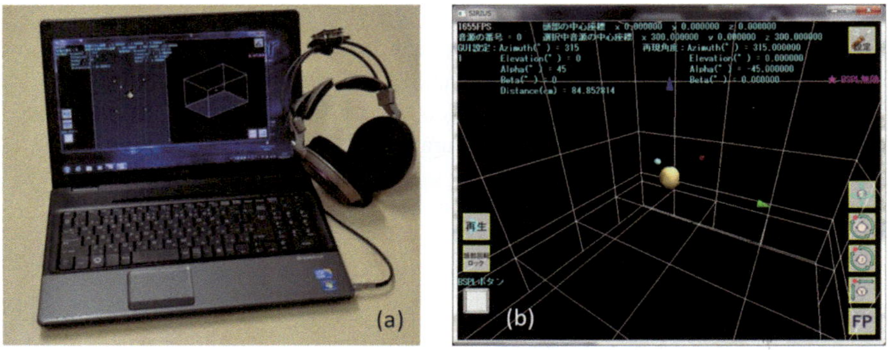

Fig. 13.2 (**a**) Photograph of system and (**b**) GUI of three-dimensional auditory display

Table 13.1 External specifications of the SIRIUS acoustic VR system

Programming language	C++, C#, MATLAB
OS	Windows 10
CPU	Core i3 2.13 GHz
Head-tracker	Acceleration (3 axes) + angular velocity (3 axes)
HRTFs	1) measured HRTFs
	2) measured median plane HRTFs + ITD
	3) parametric median plane HRTFs + ITD
Azimuth (resolution)	0° to 360° (< 1°)
Vertical angle (resolution)	−90° to +90° (< 1°)
Individualization of HRTF	Selection from the minimal parametric HRTF database
Distance of a sound image	Control based on BSPL
Maximum number of sound sources	7
System delay	< 21 ms

By using a head tracker and a three-dimensional position sensor, the direction and position of the listener's head are captured, and these changes are reflected in the signal processing in real time.

The sound image direction can be controlled for the entire sky (azimuth angle: 0° to 360°, vertical angle: −90° to +90°), and the sound image distance can also be controlled based on the BSPL (Sect. 7.2.3). By performing convolution using the overlap-add method (Sect. 10.3.2), a CPU, the clock frequency of which is 2.13 GHz, can simultaneously process up to seven sound sources. The system delay is approximately 21 ms.

13.2 Signal Processing Flow

The signal processing flow is shown in Fig. 13.3. After starting the program, initialization is performed, and the processing enters the main loop, which performs the following processes:

(1) Acquire information on the direction of the head and the position of the listener from the head tracker and three-dimensional position sensor.
(2) Calculate the relative angle and distance between the sound source and the listener based on the sound source position and the listener position set by the GUI (mouse) and the head direction.

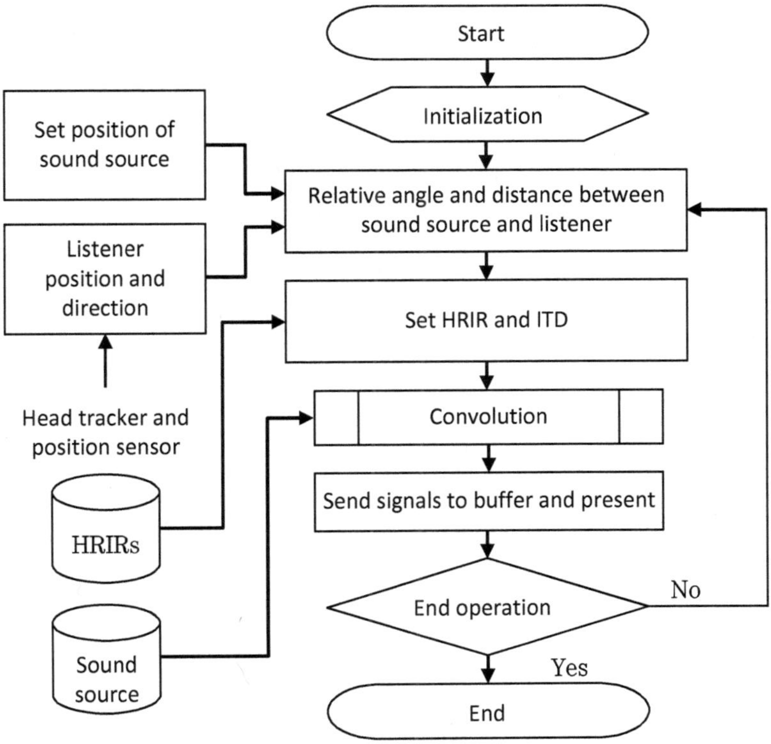

Fig. 13.3 Signal processing flow of SIRIUS

(3) Select the HRIRs corresponding to the vertical angle calculated in step (2) from the database. Then, the ITD corresponding to the lateral angle is added to the HRIRs.
(4) Perform convolution of a sound source signal and HRIRs by the overlap-add method described in Sect. 10.3.2.
(5) Send the signals calculated in step (4) to the buffer. Then, present these signals through headphones.

13.3 Application to Concert Hall Acoustics

Acoustic VR for a concert hall can be achieved by convolution of a sound source and the reflections in addition to a direct sound.

A binaural room impulse response (BRIR, see Appendixes 2 and 4) is generated by convolution of the HRIRs and the room impulse response (RIR) calculated by geometric acoustic simulation beforehand. Convolution of the BRIR and a source signal generates ear-input signals to be heard in the concert hall.

However, since the BRIR has a long response length, convolution in real time is difficult even using the overlap addition method. Therefore, a high-speed convolution algorithm using the frame division method was developed in the author's laboratory.

Figures 13.4 and 13.5 show a conceptual diagram and a flowchart, respectively, of the fast convolution algorithm. The process is as follows.

(1) Cut out the source signal every frame length (512 samples). Let the cut out blocks be denoted as $S_1, S_2, \ldots, S_n$.

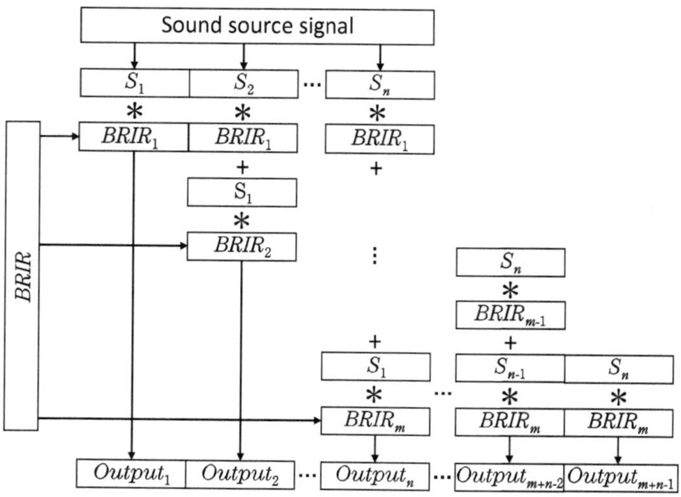

Fig. 13.4 Conceptual diagram of fast convolution algorithm using frame division

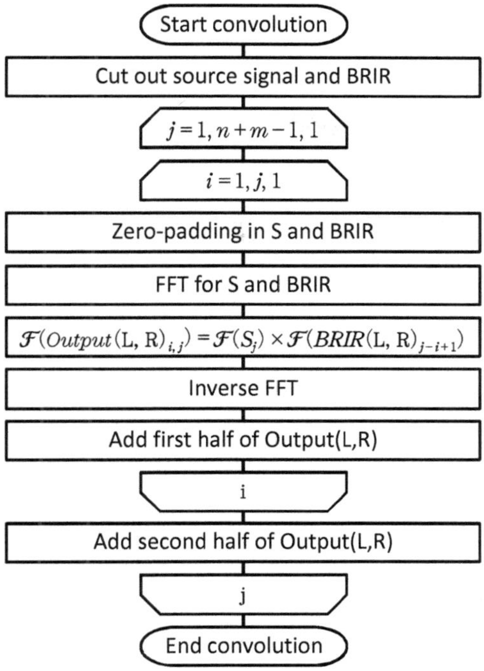

Fig. 13.5 Flow chart of fast convolution algorithm

(2) Cut out BRIR(L, R) for each frame length in the same manner. Let the cut out blocks be denoted as BRIR(L, R)$_1$, BRIR(L, R)$_2$, . . ., BRIR(L, R)$_m$.
(3) In order to obtain the output signal, Output(L, R)$_j$, of the j-th frame, calculation is performed according to Eq. (13.1).

$$Output(\mathrm{L,\ \ R})_j = \sum_{i=1}^{j} S_i * BRIR(\mathrm{L,\ \ R})_{j-i+1} \tag{13.1}$$

(4) The overlap-add method is used for the convolution of Eq. (13.1). In order to obtain the output signal, Output(L, R)$_j$, of the j-th frame according to the algorithm of the overlap-add method, the following processing is performed.

   (a) Zeros are added in each block (S$_i$, BRIR(L)$_{j-i-1}$ and BRIR(R)$_{j-i-1}$) so that the block is twice as long as the frame length.
   (b) An FFT is performed for each block.
   (c) Perform complex multiplication of S$_i$ and BRIR(L)$_{j-i-1}$, and S$_i$ and BRIR(R)$_{j-i-1}$, respectively.
   (d) An inverse FFT is performed for each of L and R of the signal obtained in c).
   (e) Add each of L and R of the signal obtained in d) to a temporary array (1024 samples).
   (f) The process from a) to e) is repeated, while i is increased until i = j.

(g) Add the first half (for 512 samples) of the signal obtained in f) and the second half (for 512 samples) of the signal obtained by the calculation of the previous frame (Output(L, R)$_{j-1}$).
(h) The first half (for 512 samples) of the signal obtained in g) is sent to the reproduction buffer. The output signal for reproduction is obtained by repeated calculation using the above processes.

Using such an algorithm, the CPU (core i7, 2.7 GHz, four-core, eight-thread) installed on a Windows 7 laptop PC is operated in six-thread parallel processing using OpenMP. The maximum response length of the BRIR was confirmed to be 92,160 samples (1.92 s), for which real time convolution is achieved.

13.4 Application to a Public Address System

Since an outdoor public address system simultaneously emits sound from loudspeakers installed at multiple points, multiple voices often arrive while overlapping listening points. The time difference of the incident sound is often on the order of hundreds of milliseconds or seconds, and the subsequent incident sound becomes a long-pass echo that reduces the speech intelligibility (see Chap. 8 and Appendix A.3).

The speech intelligibility can be evaluated directly if a speech delivered by an existing or designed outdoor public address system can be auralized. In order to simulate the intelligibility of speech accurately, it is necessary to reproduce not only the time characteristic and frequency characteristic of the incident sound but also the spatial characteristic, i.e., the three-dimensional incidence direction.

An example of simulation of the word intelligibility of an outdoor emergency public address system with a three-dimensional acoustic VR system developed in the author's laboratory is shown below. Four types of sound field, which modeled existing outdoor emergency public address systems in Tokyo, were simulated (Fig. 13.6). The sound source was a quadruple word, in which four-mora Japanese words were connected at four 1.5-mora (281.25-ms) intervals. Background noise was presented from $\pm 45°$ and $\pm$ 135° with a time delay.

Four combinations of HRTFs and headphones, as shown in Table 13.2, were used. Here, the subject's own HRTFs is denoted as "own", and the best-matching HRTF (see Sect. 4.4.1) is denoted as "bm". The HRTFs of the dummy head (B&K, Type 4128C) are denoted as "HATS". Moreover, "FEC" indicates the FEC headphones (AKG, K1000) described in Sect. 12.1, and "OPEN" denotes commercially available open-air headphones (audio-technica, ATH-AD700).

For comparison, loudspeakers were installed in an anechoic chamber, and the four types of sound fields were reproduced as an original sound field (Fig. 13.7).

Table 13.3 shows the simulation results for word intelligibility. The simulated word intelligibility has the same tendency as reproduction using loudspeakers in an anechoic room, i.e., sound field 2 < sound field 3 $\cong$ sound field 4 < sound field 1. The

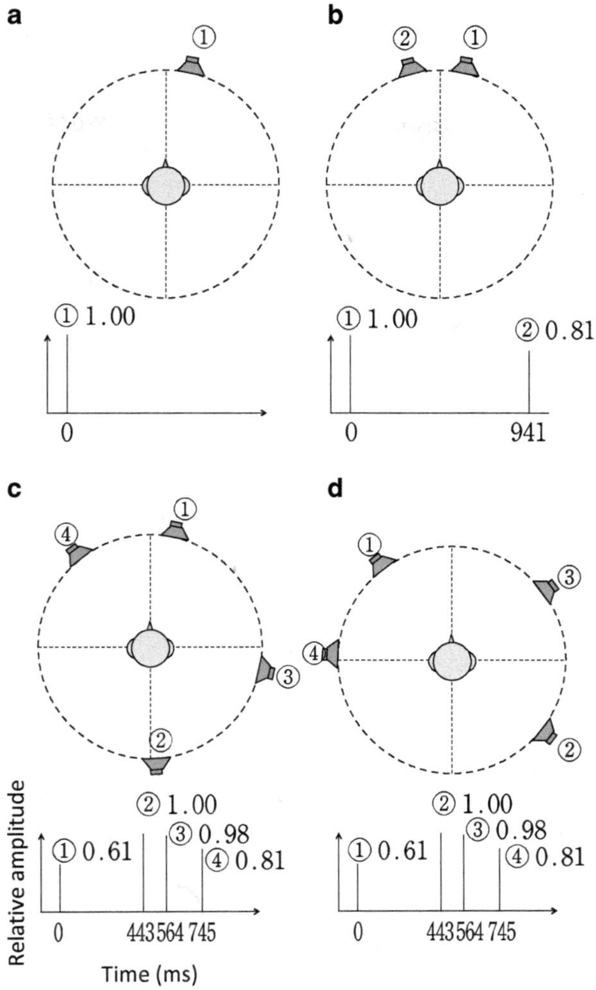

Fig. 13.6 Spatial and temporal structures of four sound fields to be simulated. (**a**) sound field 1, (**b**) sound field 2, (**c**) sound field 3, and (**d**) sound field 4

Table 13.2 Combinations of HRTFs and headphones used for auralization

Method	HRTF	Headphones
1	Subject's own (own)	FEC
2	Subject's own (own)	Open
3	Best-matching (bm)	Open
4	Dummy head (HATS)	Open

results of a chi-square test suggest that own_FEC and bm_OPEN can simulate word intelligibility with an accuracy that does not differ statistically significant from anechoic chamber reproduction.

However, the simulated word intelligibility was slightly higher than that of anechoic room reproduction for all combinations of HRTFs and headphones.

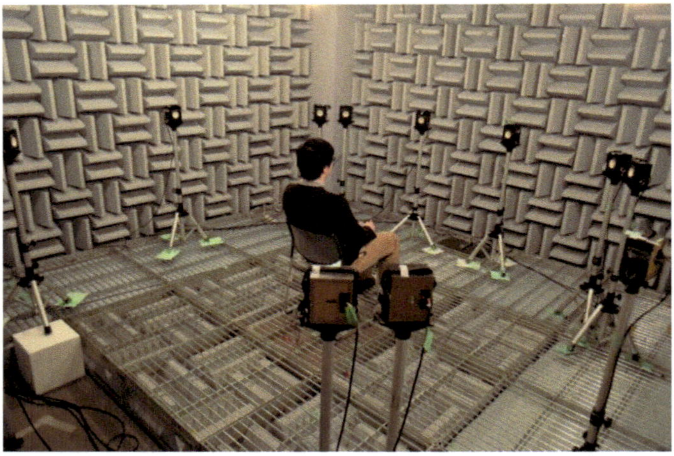

Fig. 13.7 Reproduction of four types of original sound field in anechoic chamber

Table 13.3 Comparison of word intelligibility between the original and the simulated sound field

Sound field number	Original	Simulated			
		own_FEC	own_OPEN	bm_OPEN	HATS_OPEN
1	0.77	0.80	0.83	0.81	0.84*
2	0.46	0.51	0.54*	0.47	0.50
3	0.65	0.66	0.70	0.70	0.71
4	0.64	0.65	0.71	0.70	0.74**

** and * indicate that there exist statistically significant differences between the original sound field and the simulated sound field with significance levels of 1% and 5%, respectively

13.5 Application to Searching for a Sound Source Direction

In addition to three-dimensional reproduction and presentation of sounds, systems have also been developed that use HRTFs to search for sound source directions.

For example, a method of estimating the sound source direction focusing on the interaural phase difference obtained from the input signals to the left and right ears has been proposed (Nakashima et al. 2003, Chisaki et al. 2008). This method divides the ear-input signals into multiple frequency bands and calculates the interaural phase difference in each band. Then, the intersection point of the cones (see Sect. 2.4), which are caused by the phase difference, indicates the sound source direction.

Figure 13.8 shows the estimated directions using this method for the case in which two sound sources (S1 and S2) exist. The vertical angle and azimuth angle of S1 and S2 are (30° and 20°) and (−30° and − 20°), respectively. As shown in Fig. 13.8(a), the directions of the two sound sources can be estimated accurately when the two sound sources are male and female voices. However, as shown in Fig. 13.8(b), the estimation accuracy decreases slightly for the case of a male voice and white noise.

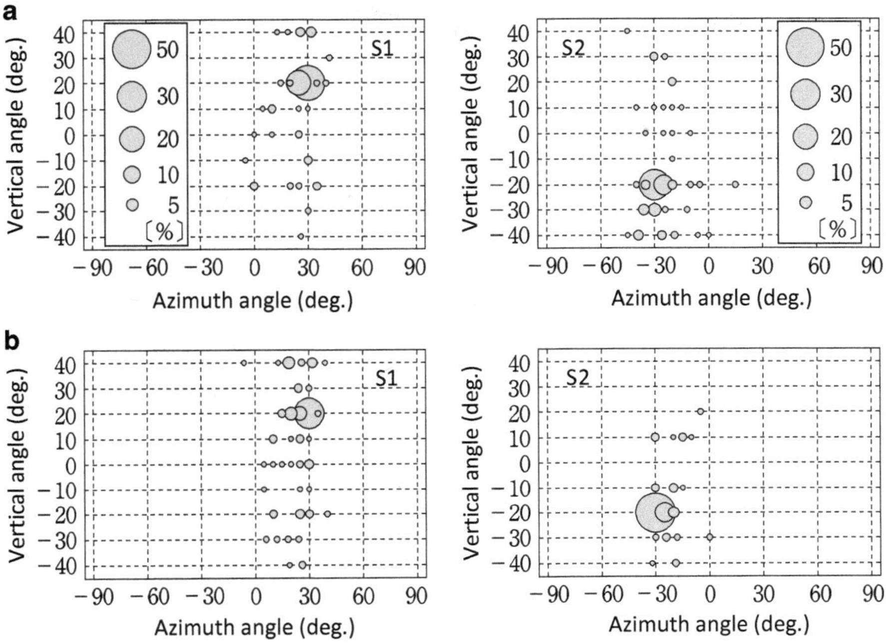

Fig. 13.8 Estimated directions for case in which two sound sources exist (Chisaki et al. 2008) (**a**) S1: Male voice, S2: Female voice (**b**) S1: Male voice, S2: White noise

A method to estimate vertical angle of the sound source by extracting the frequencies of the notches (N1 and N2) of the HRTF from sound signals recorded at both ears of a dummy head or a real head has been proposed (Iida 2010).

This method is based on the idea of using the findings that the frequencies of N1 and N2 depend strongly on the vertical angle of the sound source. This method performs signal processing as follows.

(1) Transform the input signal to the ear ipsilateral to the sound source in the time domain to spectral information by FFT.
(2) Obtain the envelope of the amplitude spectrum of the ear-input signal in order to eliminate the microscopic fluctuations using a moving average.
(3) Extract all local minima of the amplitude spectrum envelope of the input signals to the ear above 4 kHz, and set the local minima as the candidates for N1 and N2.
(4) The most probable vertical angle is estimated by collating with the relationship between the vertical angle and the frequencies of N1 and N2 (Fig. 13.9).

Figure 13.10 shows the estimated vertical angle for a sound source located in the upper median plane in 30° steps. In general, the estimation was accurate regardless of the kind of sound source. However, front-back estimation errors were observed in the cases of 0° for popular music and 30° for a female voice. This error could be related to the fact that the N1 frequency in the front direction is similar to that in the rear direction.

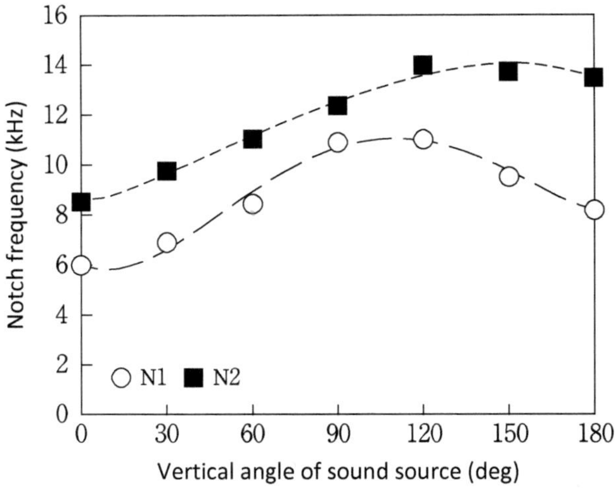

Fig. 13.9 Relationship between vertical angle of sound source and notch frequency (Iida 2010)

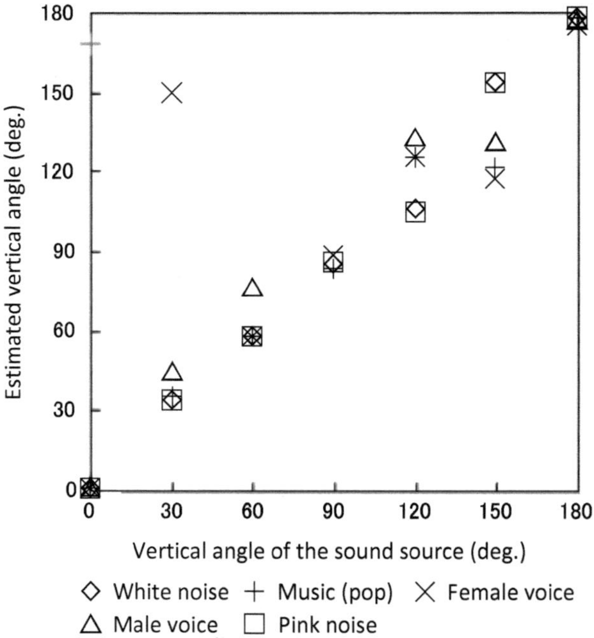

Fig. 13.10 Estimated vertical angle for sound source located in upper median plane (Iida 2010)

Furthermore, this front-back error exhibits a similar tendency to human front-back confusion. The estimated direction was behind (around 170°) for some 1-s-long parts of popular music located at vertical angle of 0° in the median plane, and to the front (around 0°) for other 1-s-long parts (Fig. 13.11). These results indicate instability in front-back estimation, which is a well-known behavior in human sound localization (Wightman and Kistler 1999).

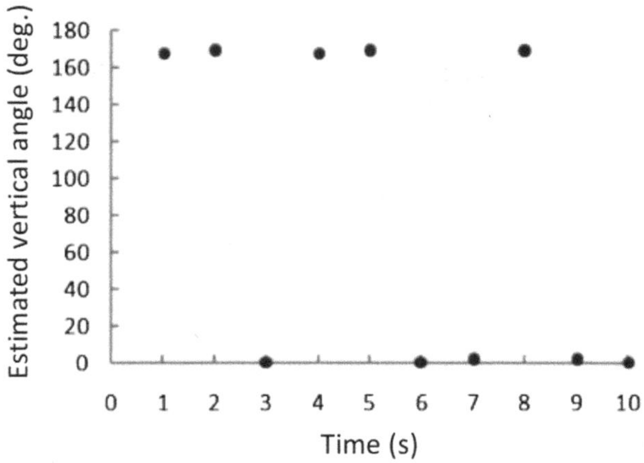

Fig. 13.11 Estimated vertical angle for various 1-s-long-parts of popular music located at vertical angle of 0° in median plane (Iida 2010)

However, as described in Chap. 4, since there are individual differences in the relationship between the vertical angle of a sound source and the N1 and N2 frequencies, it is necessary to obtain this relationship in advance for the dummy head or the real head used for recording.

As described above, although sound source direction estimation was successful in a certain range, there are problems to be solved in the future, such as the decrease of the influence of noise and reverberation. Further research is expected.

References

Chisaki Y, Kawano S, Nagata K, Matsuo K, Nakashima H, Usagawa T (2008) Azimuthal and elevation localization of two sound sources using interaural phase and level differences Acoust. Sci Tech 29:139–148

Iida K (2010) Model for estimating elevation of sound source in the median plane from ear—input signals. Acoust Sci Tech 31:191–194

Nakashima H, Chisaki Y, Usagawa T, Ebata M (2003) Frequency domain binaural model based on interaural phase and level differences. Acoust Sci Tech 24:172–178

Wightman FL, Kistler DJ (1999) Resolution of front-back ambiguity in spatial hearing by listener and source movement. J Acoust Soc Am 105:2841–2853

Yairi S, Iwaya Y, Suzuki Y (2005) Relationship betweenhead movement and total systemdelay of virtual auditory display system. IEICE Technical Report EA:2005–2038. in Japanese

Correction to: Head-Related Transfer Function and Acoustic Virtual Reality

Correction to:
K. Iida, *Head-Related Transfer Function and Acoustic Virtual Reality*, https://doi.org/10.1007/978-981-13-9745-5

The book was inadvertently published with errors in chapters 1, 5 and 7.

In chapter 1, reference citation (Morimoto et al. 1976) was added on page 3. The reference was included later as follows "Morimoto M, Joren N, Ando Y, Maekawa, Z (1976) On the Head-related Transfer Function. Technical report of technical committee of psychological and physiological acoustics, Acoust Soc Jpn H-31-1 (in Japanese)".

In chapter 5, reference citation (Morimoto et al. 2003) was added on pages 117 and 120. The reference was included later as follows "Morimoto M, Itoh M, Iida K (2003) Localization of sound image produced by two sound sources in sagittal planes. Proc. 8th Western Pacific Acoustics Conference (WESPAC VIII), TE13 (4 pages), Melbourne, Australia, 7–9 April".

Equation 7.1 was updated as $BSPL = 6\log_2\left(2^{L_l/6} + 2^{L_r/6}\right)$ on page 137.

The updated versions of the chapters can be found at
https://doi.org/10.1007/978-981-13-9745-5_1
https://doi.org/10.1007/978-981-13-9745-5_5
https://doi.org/10.1007/978-981-13-9745-5_7

K. Iida, *Head-Related Transfer Function and Acoustic Virtual Reality*,
https://doi.org/10.1007/978-981-13-9745-5_14

Appendixes

Appendix 1 Perception of Direction of an Actual Sound Source

How accurately can humans perceive the direction of an actually existing sound source (hereinafter referred to as an actual sound source), such as a loudspeaker? Here, let us consider the accuracy of sound image localization for an actual sound source.

Localization in the Horizontal Plane

Figure A.1 shows the responses of the azimuth angle of the sound image to the stimuli presented in random order from loudspeakers placed at 30° steps in the horizontal plane in an anechoic chamber. The sound source was a white noise of 200 Hz to 17 kHz, and the subjects were ten university students having normal hearing ability.

The abscissa indicates the azimuth angle of the actual sound source in the horizontal plane from 0° to 330° in 30° steps. The ordinate indicates the responded azimuth angle by the subjects. The radius of each circle is proportional to the number of responses with a resolution of 5°. The responses were distributed along a diagonal line. This indicates that the listeners localized a sound image around the direction of the actual sound source in the horizontal plane.

K. Iida, *Head-Related Transfer Function and Acoustic Virtual Reality*,
https://doi.org/10.1007/978-981-13-9745-5

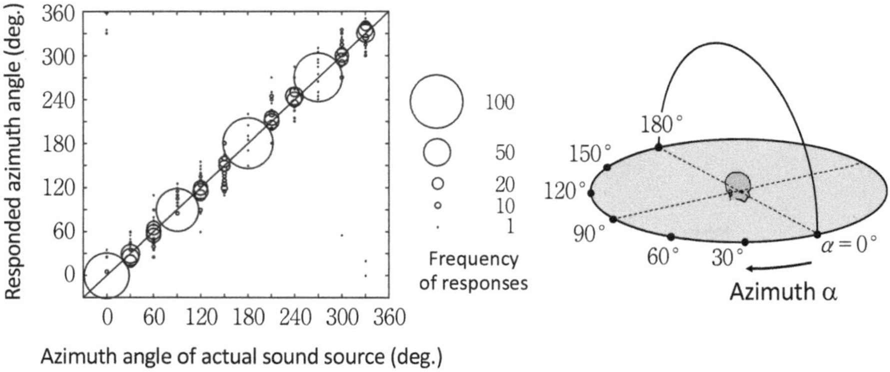

Fig. A.1 Responses to actual sound source in the horizontal plane for ten subjects

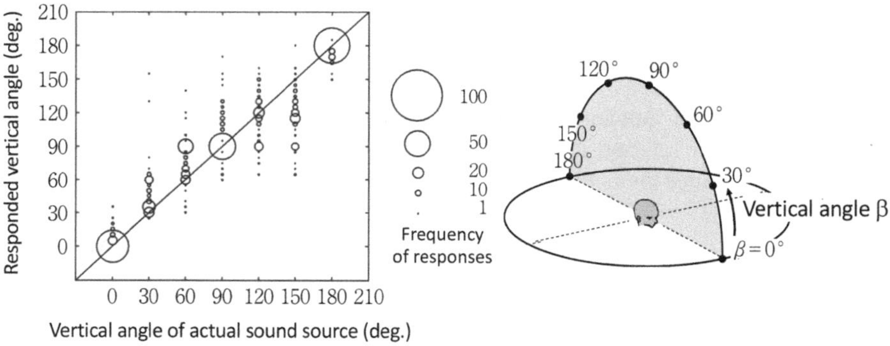

Fig. A.2 Responses to actual sound source in median plane for ten subjects

Localization in the Median Plane

Figure A.2 shows the responses for the vertical angle of the sound image to the stimuli presented in random order from loudspeakers placed at 30° steps in the upper median plane in an anechoic chamber. The sound source was white noise with a frequency of 200 Hz to 17 kHz, and the subjects were the same ten subjects used to investigate the response in the horizontal plane.

The subjects perceived a sound image in approximately the sound source direction for the sound source in the front and rear directions, while dispersion in their responses was observed for the sound source at 30° to 150°.

The author performed localization tests of an actual sound source in the upper median plane for approximately 100 subjects and found that front-back error occurred in approximately 10% of the subjects, even for an actual sound source.

Figure A.3 shows examples of three response patterns, i.e., localization with high accuracy in Fig. A.3(a), an inverted s-shaped curve in Fig. A.3(b), and front-back error in Fig. A.3(c). In Fig. A.3(c), an actual sound source of 0° is often perceived

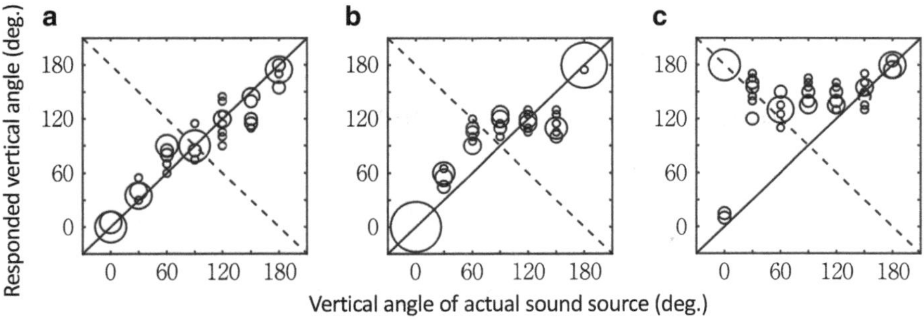

Fig. A.3 Three patterns of response to actual sound source in median plane (**a**) High accuracy (**b**) Inverted s-shaped curve (**c**) Front-back error

around 180°, and an actual sound source of 30°–90° is perceived around 120°–150°. This subject rarely perceives the sound image as upward for an actual sound source in the upper median plane.

Just Noticeable Difference in Perception of Direction

How much of a difference in the sound source direction causes a listener to perceive a difference in the sound-image direction? There have been a number of studies on the JND for direction perception for a sound source in the horizontal plane. Most of the results showed that the JND is smallest for a sound source in front, increases in the lateral direction, and then becomes small again in the rear direction.

Figure A.4 shows the JND of direction perception in the median plane, the horizontal plane, and the transverse plane for white noise (Kurosawa et al. 1982). The JND in the horizontal plane is approximately 1° for the front and rear directions and approximately 5°–10° for the lateral direction. In the median plane, the JND is approximately 5° or less for the front, 6°–25° for the upper direction, and 10° or less for the rear direction. In the transverse plane, the JND is approximately 1° for the upper direction, but increases slightly by 1°–4° for the lateral direction.

In other words, it is easier to perceive a difference in the left-right direction compared to a difference in the up-down direction for a sound source in the median plane, and it is easier to perceive a difference in the up-down direction compared to a difference in the front-back direction for a sound source in the just lateral direction.

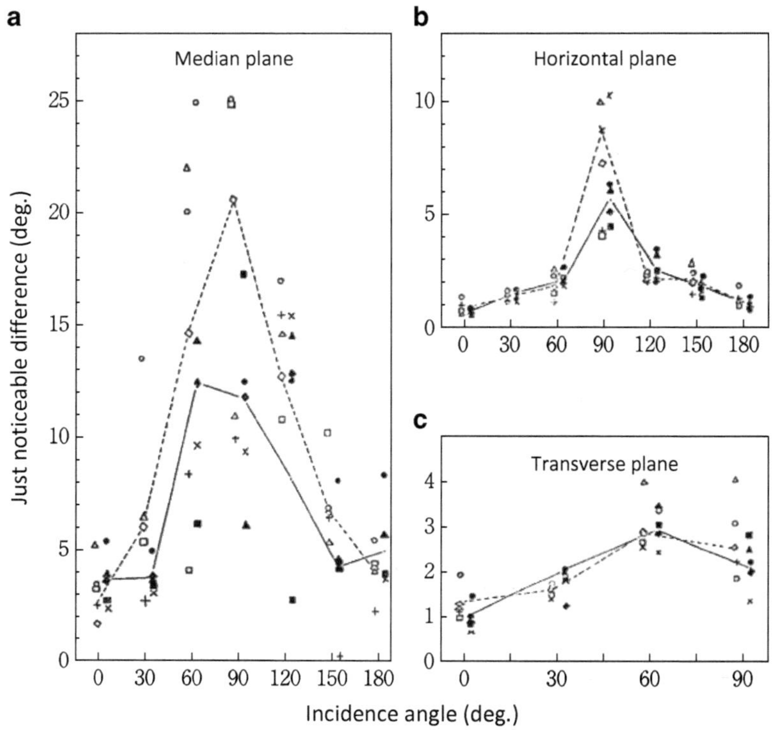

Fig. A.4 Just noticeable difference of direction perception in median plane, horizontal plane, and transverse plane for white noise. Solid and broken lines indicate the increase and decrease in angle, respectively. (Kurosawa et al. 1982)

Appendix 2 Transmission Path of Sound Waves

This appendix describes the path by which sound waves emitted from a sound source are transmitted in space and reach the eardrum of the listener.

Assuming that there exist one sound source and one listener in a certain space, the sound wave path is described by the room impulse response, the listener's HRIR, and the listener's ear canal impulse response, as shown in Fig. A.5. Each impulse response will be described in order.

Room Impulse Response

First, let us consider the transmission path for an acoustic wave from a sound source to a point (receiving point) corresponding to the center of the head without the listener.

As shown in Fig. A.6, a direct sound and reflections by way of a wall or ceiling reach the receiving point. A representation of such a transfer process in the time domain is referred to as a room impulse response (RIR). Schematically, the RIR is expressed as shown in Fig. A.7. The early reflections arrive discretely, and the late

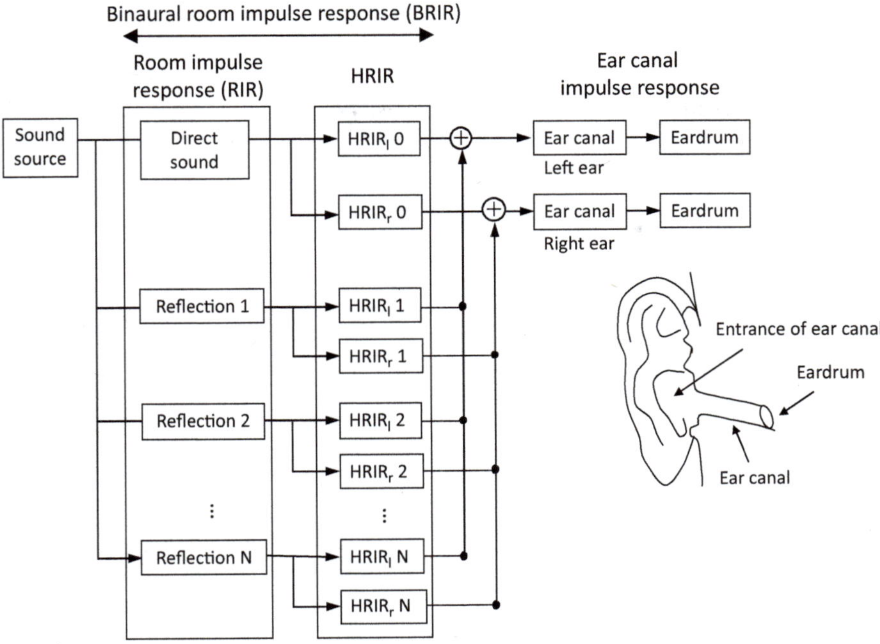

Fig. A.5 Sound transmission path from sound source to listener's eardrums

Fig. A.6 Sound transmission path from sound source to receiving point

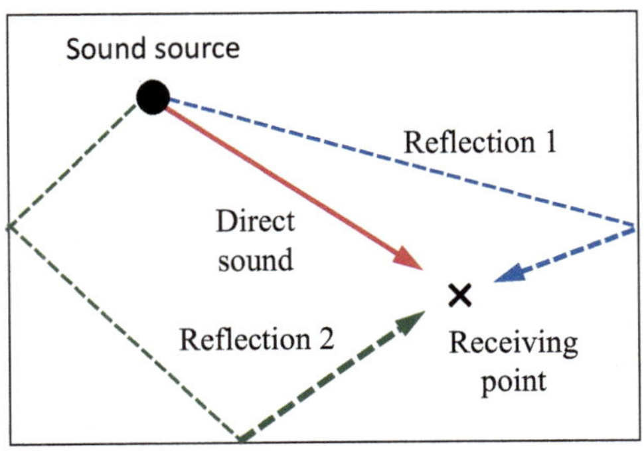

Fig. A.7 Schematic diagram of RIR

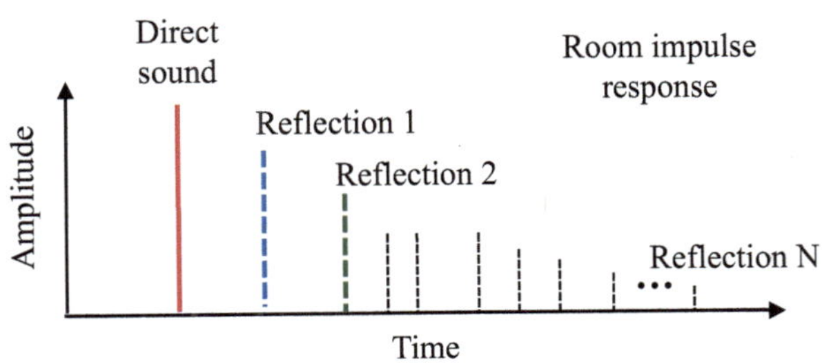

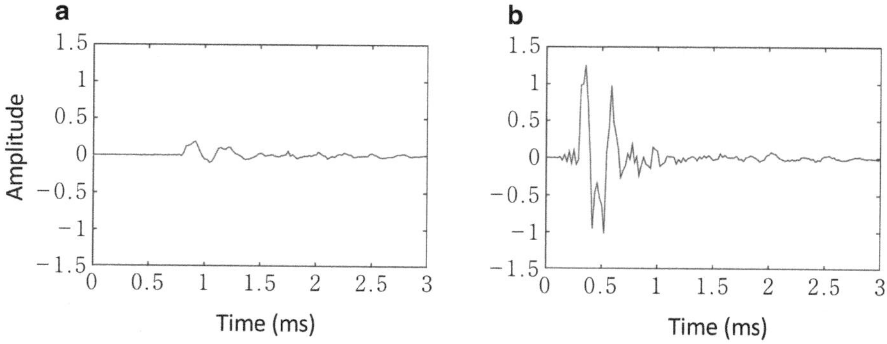

Fig. A.8 Head-related impulse response for sound wave coming from azimuth angle of 60° in horizontal plane. (**a**) left ear, (**b**) right ear

reflections (reverberation) reach the listener continuously, although the amplitudes are small.

The RIR is determined by the shape of the space, the reflection characteristics of the boundary surfaces, such as walls, a ceiling and a floor, and the positions of the sound source and the receiving point. The response length of the RIR is generally in the range of several hundred milliseconds to several seconds for a typical room.

Head-Related Impulse Response

Next, we consider the influence of the listener. The direct and reflected sounds shown in Figs. A.6 and A.7 are influenced by the listener's head and pinnae and reach the entrances of the ear canals of the left and right ears (two receiving points). Such a change in the physical characteristics of the incident wave by the head represented in the time domain is referred to as a head-related impulse response (HRIR).

Figure A.8 shows an example of a measured HRIR for a sound wave coming from an azimuth angle of 60° in the horizontal plane. The HRIR of the right ear increases earlier in time, and its amplitude is larger compared to that of the left ear. Thus, the HRIRs generate a time difference and a level difference between the sound waves reaching the left and right ears. These differences become large when sound waves come from the lateral direction. These time difference and level difference are the ITD and the ILD (see Chap. 2).

Binaural Room Impulse Response

A combination of RIR and HRIR, i.e., a time domain representation of the propagation process from the sound source to the entrance of the ear canal, is referred to as the binaural room impulse response (BRIR), which is defined for each of the left and right ears.

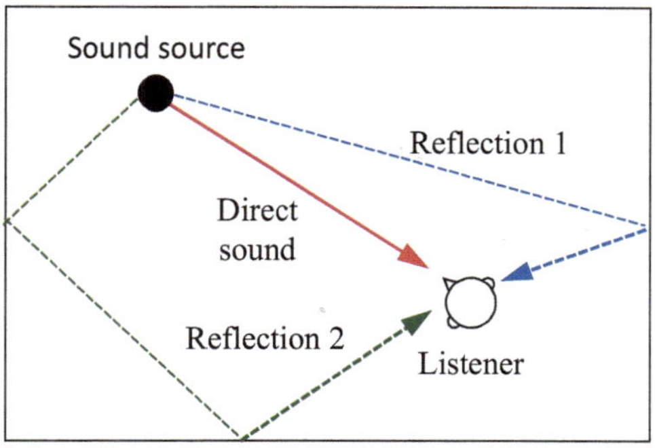

Fig. A.9 Sound transmission path from sound source to listener

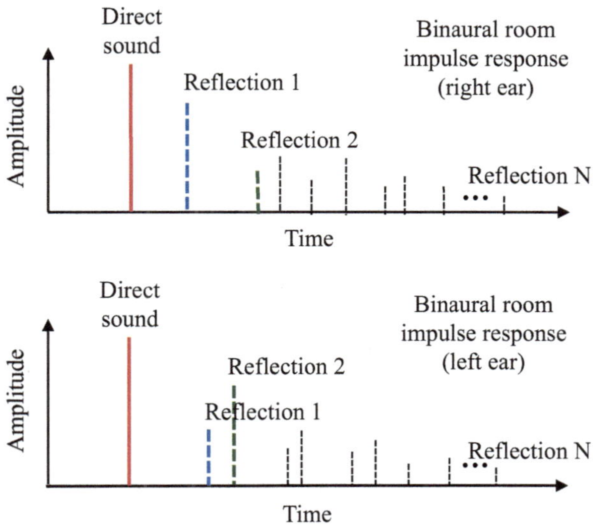

Fig. A.10 Schematic diagram of BRIR

Schematic diagrams of the transmission path and BRIRs are shown in Figs. A.9 and A.10. In this case, since the direct sound comes from the front of the listener, the sound arrives with the same sound pressure amplitude at the same time at the left and right ears. However, since Reflection 1 comes from the right side of the listener, the arrival time at the right ear is earlier and the amplitude is larger than those for the left ear. In addition, since Reflection 2 comes from the left side, its behavior is reversed as compared to Reflection 1.

Ear Canal Impulse Response

The sound waves that reach the entrances of the ear canals pass through the ear canals and reach the eardrums.

The ear canal is a tube with a diameter of 7–8 mm and a length of approximately 25 mm, and the sound field can be regarded as one-dimensional for a frequency of approximately 17 kHz or less. Therefore, the propagation characteristics in the ear canal are constant, independent of the incidence direction of the sound wave.

In addition, since the ear canal is a tube with one closed end at the eardrum, resonance occurs at a frequency at which the 1/4 wavelength matches the tube length, i.e., 3–4 kHz. As a result, the sound pressure is amplified by approximately 10 dB at this frequency regardless of the incidence direction, at the eardrum. This is P1 (see Sect. 3.5.2).

Summary of the Transmission Path

A sound wave emitted from a sound source reaches the vicinity of the listener's head as a direct sound and a number of reflections represented by the RIR. These incident waves arrive at the entrances of the left and right ear canals under the influence of the HRIRs and then pass through the ear canals to reach the eardrums of the listener.

Thus far, the sound propagation process has been explained in terms of the impulse response (time domain), but this process can also be expressed by transfer functions (frequency domain). Although the ITD and the ILD can be observed in the HRIRs, analysis of more detailed physical features using HRIRs is not easy.

Instead, its physical meaning is better expressed in the frequency domain. This is the reason why we discuss HRTFs in this book.

Appendix 3 Prediction Method of Room Acoustics

A method by which to predict whether the impulse response and physical measure of room acoustics can reach target values would be very useful in the acoustical design of a concert hall. A prediction method for a sound field is also important to prevent acoustic disturbances such as echoes.

Prediction methods for sound fields are classified roughly into two types: numerical calculation using a computer and measurement using a scale model (Vorlander 2011). These methods are outlined in the following.

Numerical Calculation

Numerical calculation methods for the sound field can be roughly divided into geometric acoustic simulation, which assumes light-like propagation without considering the wave property of sound, and wave acoustic simulation, which considers the wave property of sound.

In these methods, the shape of the room is recreated by a wire frame model, for example, and a sound absorption coefficient corresponding to the wall material is assigned to each surface in order to construct the room in a computer.

Image Method

In the image method, an imaginary sound source is determined based on the idea that when a sound wave encounters a wall surface, a reflection is emitted only in the direction symmetrical to the incidence angle with respect to the normal direction.

Therefore, the first-order imaginary sound source is found at a position (dashed line) symmetrical to the sound source with respect to each wall surface (thick line), as shown in Fig. A.11. In this two-dimensional plane example, four first-order imaginary sound sources are obtained.

Next, a first-order imaginary sound source is regarded as a sound source, and second-order imaginary sound sources are determined at symmetrical positions with respect to each wall surface. Similarly, higher-order imaginary sound sources can be obtained.

Then, assuming that a spherical wave is emitted from each imaginary sound source, the path (broken line) to the receiving point is determined. The point that intersects the wall is the reflection point.

The time-of-arrival of each incident wave is determined based on the path length. The relative sound pressure amplitude is calculated considering the distance attenuation and sound absorption coefficient of the walls (Fig. A.12).

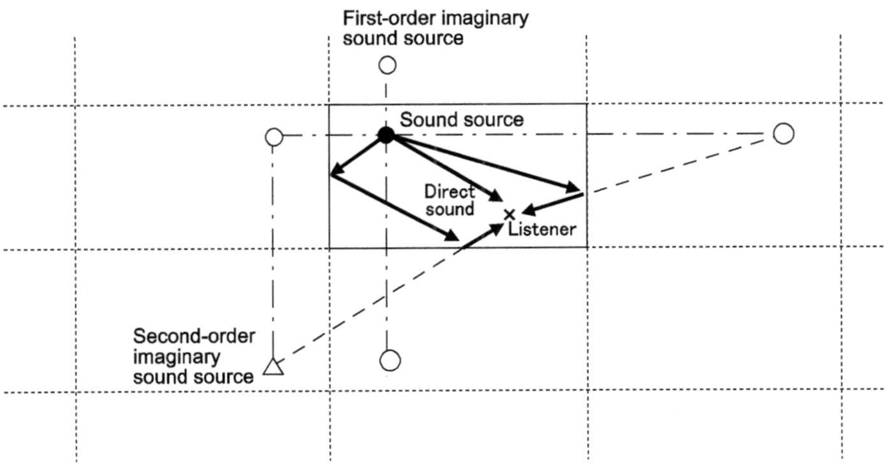

Fig. A.11 Example of imaginary sound sources on two-dimensional plane. Open circles indicate the four first-order imaginary sound sources. The open triangle indicates a second-order imaginary sound source

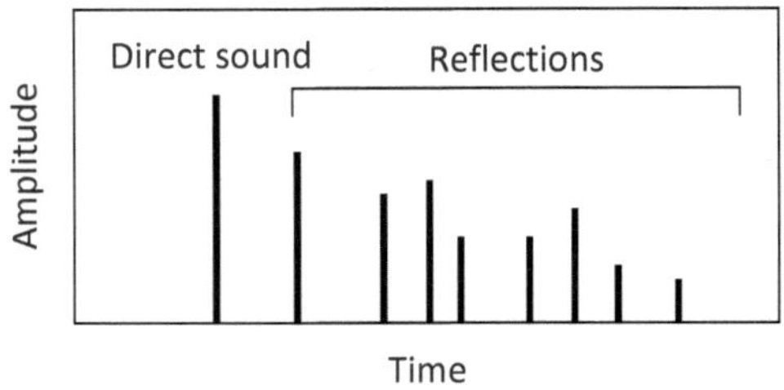

Fig. A.12 Example of temporal structure of incident waves obtained by image method

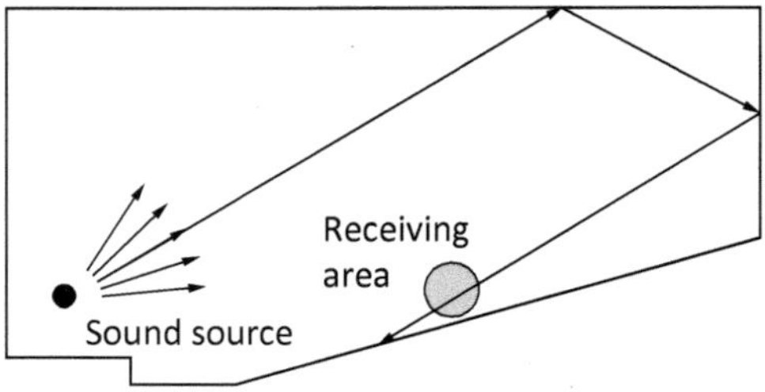

Fig. A.13 Conceptual diagram of ray tracing method

Since the image method does not consider the wave property, the accuracy can be maintained only when the wall surface is sufficiently large compared to the wavelength of the sound wave.

In other words, although the accuracy is relatively high for a high-frequency sound, an error occurs due to not considering wave effects (such as diffraction) at low frequency.

In addition, the early reflections are relatively easily obtained, although it is necessary to increase the reflection order in order to obtain the late reverberations, and there exists a problem in that the calculation time increases exponentially.

Ray Tracing Method

As shown in Fig. A.13, the ray tracing method is a method of emitting a large number of rays from a sound source at an equal solid angle and tracking the reflection path of each ray.

As in the image method, the sound ray incident on the wall surface travels in the direction symmetrical to the incidence angle with respect to the normal direction.

The longer the tracking time, the wider the sound beam interval. Therefore, a spherical sound receiving area is set instead of a sound receiving point, and sound rays passing through the receiving area are regarded as incident sounds.

In the case of the ray tracing method, the low-frequency prediction accuracy becomes a problem, as in the image method.

Compared with the image method, the ray tracing method has the advantage of being able to track later reflections, but has the disadvantage that a large number of rays must be emitted.

Numerical Calculation Method Considering the Wave Property of Sound

The finite element method (FEM) and the boundary element method (BEM) have been investigated as numerical methods for calculating the room sound field considering the wave property.

Recently, there have also been attempts to apply the finite difference time domain (FDTD) method to the calculation of a room sound field.

However, in order to handle the high-frequency region, it is necessary to model the room using a large number of very small elements.

Scale Model Experiment

A scale model based on design drawings and acoustic measurements in the model can be used to confirm the characteristics of the sound field of a room.

Assuming that the scale of the model is 1/n, the frequency in the model must to be n times that of the real room in order for the model and the real room to be physically similar. Therefore, the sound absorption coefficient of the model wall at n times the frequency must be matched to the sound absorption coefficient for the real wall at the original frequency.

Since the frequency of the sound source signal for measuring the characteristics in the model is also multiplied by n, a discharge pulse or a tweeter are used often in the measurement. In some cases, the air in the model is replaced with nitrogen in order to make the air absorption of sound similar as well.

The scale model experiment has a high prediction accuracy compared with computer simulation at the present time. However, it is difficult to respond promptly to changes in design plans and boundary conditions.

Appendix 4 Time Window

The time window is used to cut out only the necessary part from the signal in the time domain. However, the characteristics of the time window affect the analysis results for the signal, and, in addition to the main lobe, which is the original frequency component, side lobes with components not included in the original signal appear. Although it is desirable to use a time window having a low side-lobe level (wide dynamic range) and a narrow main lobe (high frequency resolution), these are generally in a trade-off relationship. In this appendix, typical time windows that have been derived are introduced.

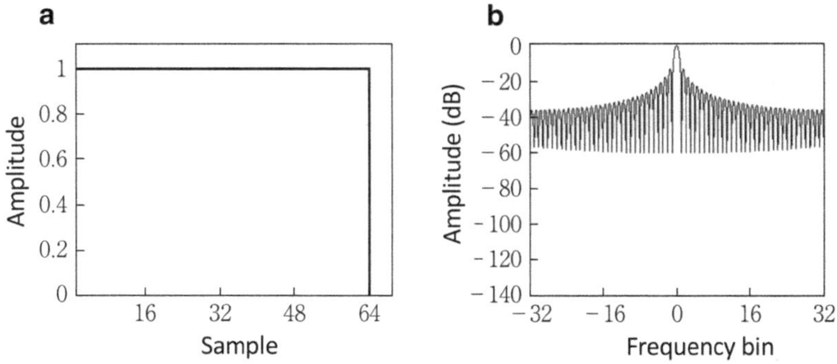

Fig. A.14 Temporal and frequency characteristics of rectangular window (**a**) Temporal characteristics (**b**) Frequency characteristics

Rectangular Window

The rectangular window is expressed by the following equation. The frequency resolution is high, while the level of the side lobes is high (Fig. A.14).

$$w(n) = \begin{cases} 1 & 0 \leqq n \leqq N-1 \\ 0 & \text{Others} \end{cases} \tag{A.9}$$

Hanning Window

The Hanning window is one of the most commonly used time windows (Fig. A.15).

$$w(n) = \begin{cases} 0.5 - 0.5\cos\left(\frac{2\pi n}{N}\right) & 0 \leqq n \leqq N-1 \\ 0 & \text{Others} \end{cases} \tag{A.10}$$

Hamming Window

The Hamming window is one of the most commonly used window functions. The frequency resolution of the Hamming window is higher than that of the Hanning window, while its dynamic range is narrow. The pressure amplitude is discontinuous at both ends of the window (Fig. A.16).

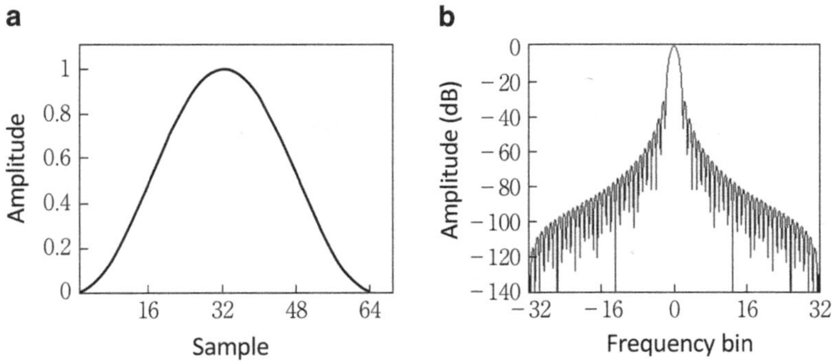

Fig. A.15 Temporal and frequency characteristics of Hanning window (**a**) Temporal characteristics (**b**) Frequency characteristics

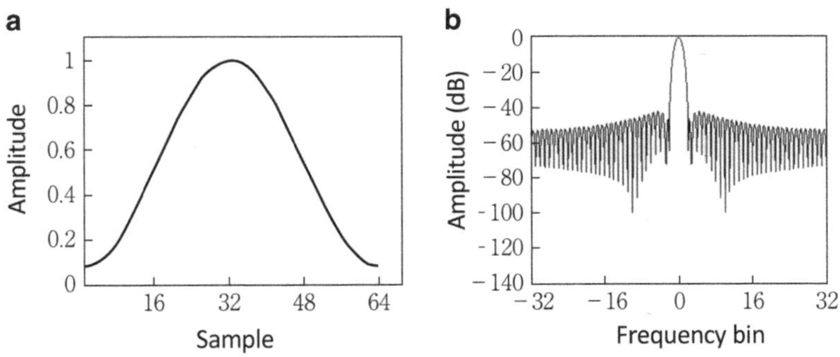

Fig. A.16 Temporal and frequency characteristics of Hamming window (**a**) Temporal characteristics (**b**) Frequency characteristics

$$w(n) = \begin{cases} 0.5 - 0.46\cos\left(\dfrac{2\pi n}{N}\right) & 0 \leqq n \leqq N-1 \\ 0 & \text{Others} \end{cases} \tag{A.11}$$

Blackman Window

The Blackman window has worse frequency resolution than the Hanning window or the Hamming window but has a wide dynamic range (Fig. A.17).

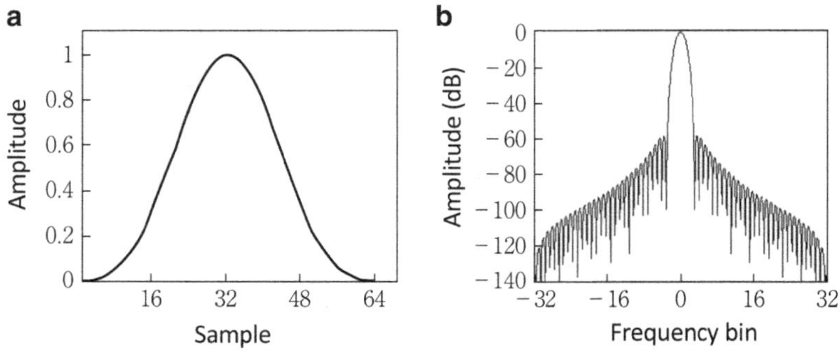

Fig. A.17 Temporal and frequency characteristics of Blackman window (**a**) Temporal characteristics (**b**) Frequency characteristics

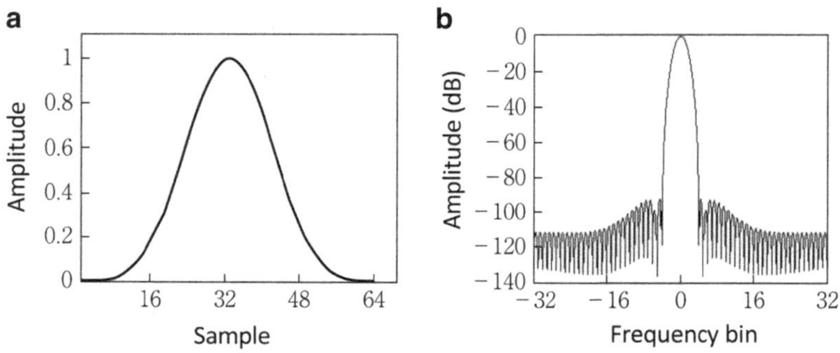

Fig. A.18 Temporal and frequency characteristics of Blackman-Harris window (**a**) Temporal characteristics (**b**) Frequency characteristics

$$w(n) = \begin{cases} 0.42 - 0.5\cos\left(\frac{2\pi n}{N}\right) + 0.08\cos\left(\frac{4\pi n}{N}\right) & \left(0 \leqq n \leqq N-1\right) \\ 0 & \text{Others} \end{cases} \tag{A.12}$$

Blackman-Harris Window

The Blackman-Harris window has minimal side lobes and a wide dynamic range (Fig. A.18).

```
clear
N = 2^11 ; // window length
N=N;
//----- rectangular -----//
for i=1:N
      RECTANGULAR(i) = 1;
end

//-----    Hanning     -----//
if modulo(N,2)==0
      for i=1:N
          HANNING(i) = 0.5-0.5*cos(2*%pi*(i-1)/(N-1));
      end
else
      for i=1:N
          HANNING(i) = 0.5-0.5*cos(2*%pi*(i-0.5)/(N-1));
      end
end

//-----    Hamming     -----//
if modulo(N,2)==0
      for i=1:N
          HAMMING(i) = 0.54-0.46*cos(2*%pi*(i-1)/(N-1));
      end
else
      for i=1:N
          HAMMING(i) = 0.54-0.46*cos(2*%pi*(i-0.5)/(N-1));
      end
end
//-----     Blackman      -----//
if modulo(N,2)==0
      for i=1:N
          BLACKMAN(i)=0.42-0.5*cos(2*%pi*(i-1)/
          (N-1))+0.08*cos(4*%pi*(i-1)/(N-1));
      end
else
      for i=1:N
          BLACKMAN(i)=0.42-0.5*cos(2*%pi*(i-0.5)/
          (N-1))+0.08*cos(4*%pi*(i-0.5)/(N-1));
      end
end

//-----     Blackman-Harris      -----//
if modulo(N,2)==0
      for i=1:N
          BLACKMAN_HARRIS(i)=0.35875-
0.48829*cos(2*%pi*(i-1)/
(N-1))+0.14128*cos(4*%pi*(i-1)/(N-1))-0.01168*cos(6*%pi*(i-1)/
(N-1));
      end
```

Fig. A.19 Sample program for window processing

```
else
      for i=1:N
          BLACKMAN_HARRIS(i)=0.35875-0.48829*cos(2*%pi*(i-0.5)/
(N-1))+0.14128*cos(4*%pi*(i-0.5)/(N-1))-
0.01168*cos(6*%pi*(i-0.5)/
(N-1));
      end
end

// Read sound source //
[x,fs] = wavread('music.wav');
X = fft(x);
XdB=20*log10(abs(X));

// window processing //
for i = 1:N
x1(i) = x(i) .* RECTANGULAR(i);
x2(i) = x(i) .* HANNING(i);
x3(i) = x(i) .* HAMMING(i);
x4(i) = x(i) .* BLACKMAN(i);
x5(i) = x(i) .* BLACKMAN_HARRIS(i);
end


// amplitude spectrum //
X1 = fft(x1);
X2 = fft(x2);
X3 = fft(x3);
X4 = fft(x4);
X5 = fft(x5);
X1dB = 20*log10(abs(X1));
X2dB = 20*log10(abs(X2));
X3dB = 20*log10(abs(X3));
X4dB = 20*log10(abs(X4));
X5dB = 20*log10(abs(X5));
ff = fs.*[0:N/2-1]/N
// plot //
clf
subplot(321);plot2d(fs*[0:(length(x)/2-1)]/
length(x),XdB(1:length(x)/2));
xlabel( 'Freq (Hz)' );ylabel( 'Amp. (dB)' );square(0,-80,fs/2,60);
subplot(322);plot2d(ff,X1dB(1:N/2));
xlabel( 'Freq (Hz)' );ylabel( 'Amp. (dB)' );square(0,-80,fs/2,60);
subplot(323);plot2d(ff,X2dB(1:N/2));
xlabel( 'Freq (Hz)' );ylabel( 'Amp. (dB)' );square(0,-80,fs/2,60);
subplot(324);plot2d(ff,X3dB(1:N/2));
xlabel( 'Freq (Hz)' );ylabel( 'Amp. (dB)' );square(0,-80,fs/2,60);
subplot(325);plot2d(ff,X4dB(1:N/2));
xlabel( 'Freq (Hz)' );ylabel( 'Amp. (dB)' );square(0,-80,fs/2,60);
subplot(326);plot2d(ff,X5dB(1:N/2));
xlabel( 'Freq (Hz)' );ylabel( 'Amp. (dB)' );square(0,-80,fs/2,60);
```

Fig. A.19 (continued)

$$w(n) = \begin{cases} 0.35875 - 0.48829\cos\left(\frac{2\pi n}{N}\right) + 0.14128\cos\left(\frac{4\pi n}{N}\right) \\ -0.01168\cos\left(\frac{6\pi n}{N}\right) \quad (0 \leqq n \leqq N-1) \\ 0 \quad \text{Others} \end{cases} \tag{A.13}$$

A sample program (Scilab) that cuts off the signal with the time window described above is shown in Fig. A.19.

Appendix 5 Method for Making an Earplug-Type Microphone

The most important instrument for measuring the HRTFs is a microphone. In terms of safety, accuracy, and theoretical background, it is appropriate to measure the HRTFs at the entrance of the blocked ear canal using earplug-type microphones. In this appendix, I will describe how to make earplug-type microphones and their use. The author has created earplug-type microphones in over 150 cases in the laboratory. However, readers are encouraged to take full responsibility for the subject's safety (protection of the ear canals and eardrums).

Making an Ear Mold

The subject's ear mold is obtained to create an earplug-type microphone, which fits snugly into the entrance of the ear canal.

Materials and Equipment

1. Reverse ear mold sampling tools for custom-made hearing aids: silicone, silicone injection syringes, earplugs, and an earplug insertion rod with a light.
2. Ear mold making tools: fine plaster (e.g., for dental use) and a vibrator.
3. Other tools: scale (0.1 g accuracy), measuring cylinder or measuring cup, dropper, rubber bowl, mixing rod, paper cup, and cotton swab.

Procedure for Making a Reverse Ear Mold

A reverse ear mold of the subject is made according to the following procedure.

1. Clean the subject's ears.

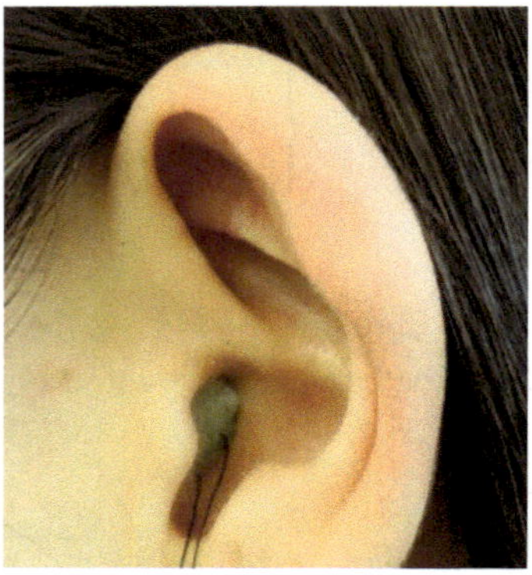

Fig. A.20 Pinna and earplug with thread for extraction

Fig. A.21 Two types of silicone to be mixed

2. In order to prevent the silicone from reaching the subject's eardrum, place an earplug with a thread for extraction (Fig. A.20) in the ear canal (approximately 5–7 mm from the entrance).
3. Mix two types of silicone (Fig. A.21) in a 1:1 ratio by hand kneading for quick uniformity.
4. Roll around 1/3 of the silicone and place it into the syringe.
5. Inject the silicone all at once into the pinna (Fig. A.22). The task is easier if the subject's head is lying on a desk.
6. Cover the entire pinna with the remaining silicone (Fig. A.23). Extending the silicone to the back of the ear and sideburns will make it easier to obtain a final mold with gypsum later.
7. Wait for the silicone to set (approximately 5–7 min, depending on room temperature).
8. Confirm that the silicone has solidified and then remove it (Fig. A.24). At this time, gently remove the mold from the front, slowly allowing air in (Be aware that there is a risk of damaging the eardrum if the mold is removed suddenly).

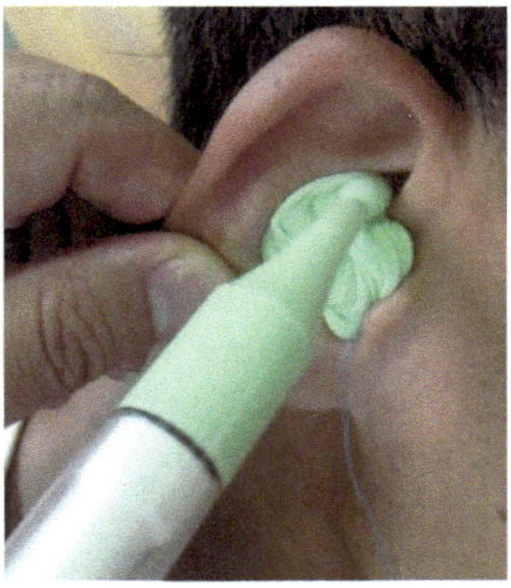

Fig. A.22 Injection of silicone into pinna

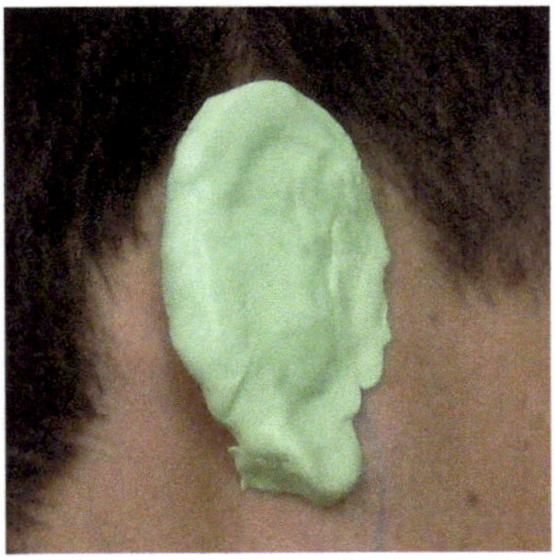

Fig. A.23 Pinna covered with silicone

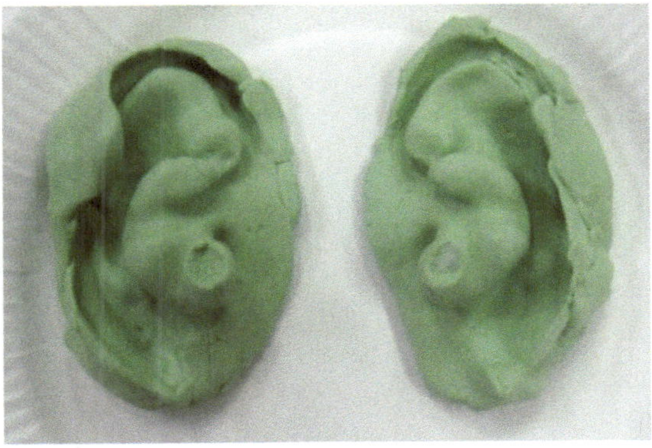

Fig. A.24 Removed reverse ear mold

Procedure for Making the Final Mold

Next, plaster is poured into the inverse ear mold to make the final mold.

1. Mix plaster and water (Fig. A.25) quickly for approximately 30–60 s. Mix the plaster and water gently but quickly (Fig. A.26), which is easier using a rubber bowl.
2. Pour the plaster into a reverse ear mold made of silicone. Adjust the shape of the plaster so that there are no air gaps (Fig. A.27).
3. Vibration is applied to the reverse mold with a vibrator and plaster is injected in the gap (Figs. A.28 and A.29).
4. Firmly cover the reverse ear mold with the remaining plaster (Fig. A.30).
5. Allow the mixture to harden for 10–20 min. Before removal, allow the heated plaster to cool.
6. Remove the silicone to obtain the final ear mold (Figs. A.31 and A.32).

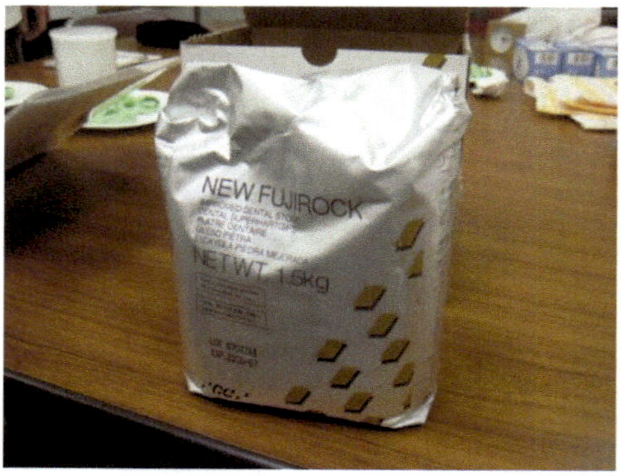

Fig. A.25 Example of fine plaster

Fig. A.26 Mixing plaster and water using rubber bowl

Fig. A.27 Pour plaster into reverse ear mold

Fig. A.28 Photograph of vibrator

Fig. A.29 Injection of plaster into gap

Fig. A.30 Reverse ear mold covered by plaster

Fig. A.31 Removing silicone from plaster

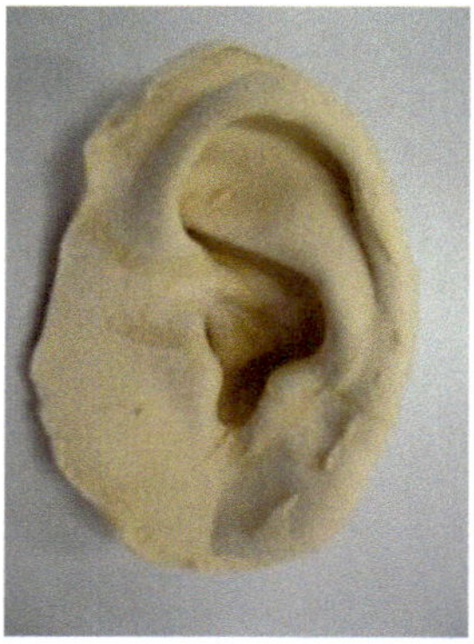

Fig. A.32 Final ear mold

Making a Microphone

Materials and Equipment

The materials and equipment needed to make an earplug-type microphone are listed below.

1. Material: microphone unit (e.g., WM64AT102 (Panasonic), FG3329 (Knowles), lead wire, stereo mini-plug, silicone, solder, and mold release agent.
2. Equipment: soldering iron, reverse tweezers, and precision screwdriver.

Preparation of a Microphone

Attach the lead wire to the microphone unit.

1. Cut the lead wire to a length of approximately 30–40 cm.
2. Cut the vinyl at the tip of the lead wire 1–2 mm and plate the tip with solder.
3. Solder the lead wire to the microphone unit (Figs. A.33 and A.34). At this time, using reverse tweezers will make it easier to fix the microphone unit. Apply solder within 1 s, as overheating will break the microphone unit.
4. Solder the other end of the lead wire to a stereo mini-plug (Fig. A.35).
5. Connect it to a microphone amplifier and confirm that the microphone works.

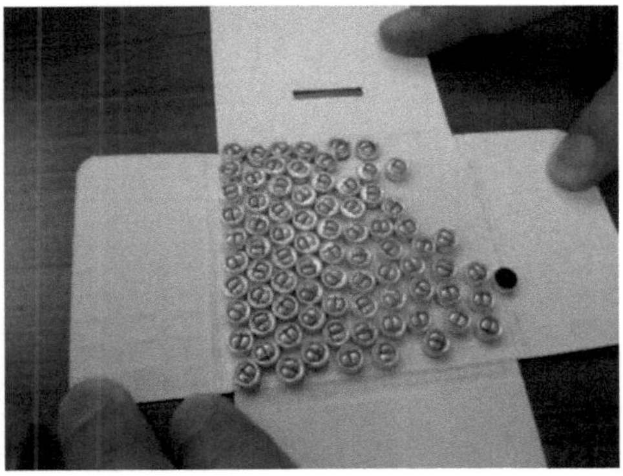

Fig. A.33 Microphone units (WM64AT102, Panasonic)

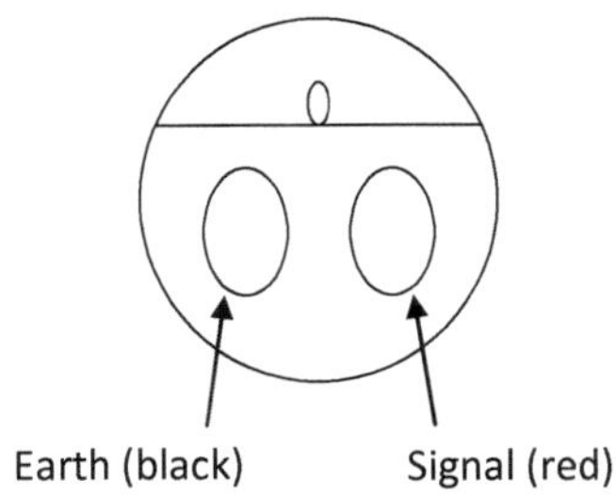

Fig. A.34 Microphone terminals

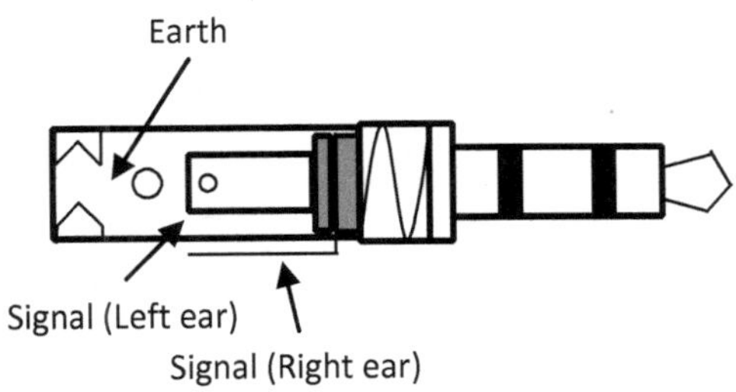

Fig. A.35 Terminals of stereo mini-plug

Making an Earplug-Type Microphone

Make an earplug-type microphone using an ear model and the microphone described above.

1. Apply a mold release agent to the ear mold, so as to make it easy to take out the silicone, which is to be poured later.
2. After the release agent dries, determine the position of the microphone. Adjust the microphone unit's diaphragm to the entrance of the ear canal.
3. Mix silicone (blue mix) in a 1:1 ratio.
4. Transfer the silicone to a paper cup, for example, and pour the silicone into the ear mold (Fig. A.36). Narrowing the tip of the paper cup will make it easier to pour the silicone into the ear mold.
5. Wait for approximately 10 min until the silicone hardens.
6. Carefully remove the entire piece of silicone using a precision screwdriver (Fig. A.37).

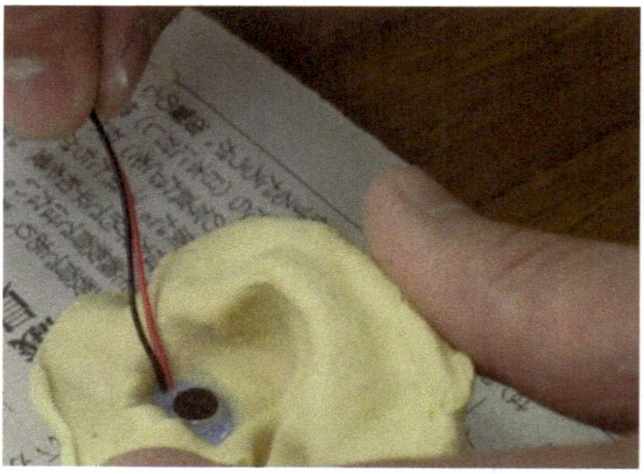

Fig. A.36 Pouring silicone into ear mold

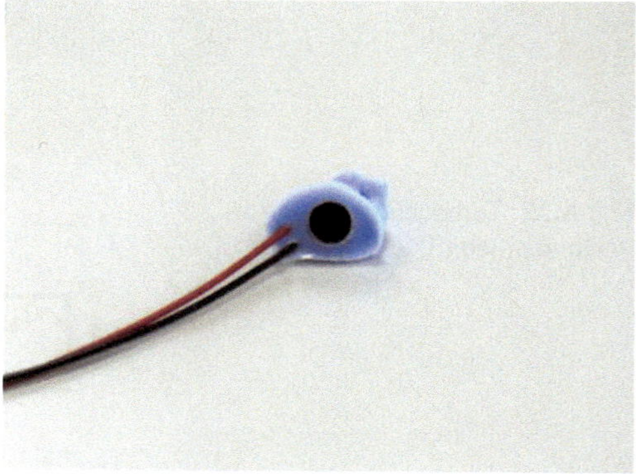

Fig. A.37 Earplug-type microphone

Appendix 6 HRTFs Using 96-kHz Sampling

Due to the audible range of humans, HRTFs were often measured using 48-kHz sampling. However, sound source signals with 96-kHz and 192-kHz sampling have appeared, and it has become necessary to prepare corresponding HRTFs.

An example of an HRTF measured using a swept-sine signal with 96-kHz sampling and a tweeter (Fostex, FT28D) is shown in Fig. A.38. The solid line and broken line denote the HRTF with 96-kHz sampling and 48-kHz sampling, respectively. The amplitude of the HRTF does not fall off in the frequency range higher than 24 kHz. A large number of notches and peaks are observed.

Figure A.39 shows the amplitude spectra of the HRTFs with 96-kHz sampling at the right ear for sound sources in the horizontal plane (0°–330°, 30° steps). As in the frontal direction, the amplitude of the HRTF does not fall off in the frequency range higher than 24 kHz, and notches and peaks are observed.

Therefore, when trying to reproduce three-dimensional sound using a 96-kHz sampling sound source (aside from whether or not this component is audible), from a physical point of view, it is reasonable to use HRTFs, which contain components in the frequency range above 24 kHz.

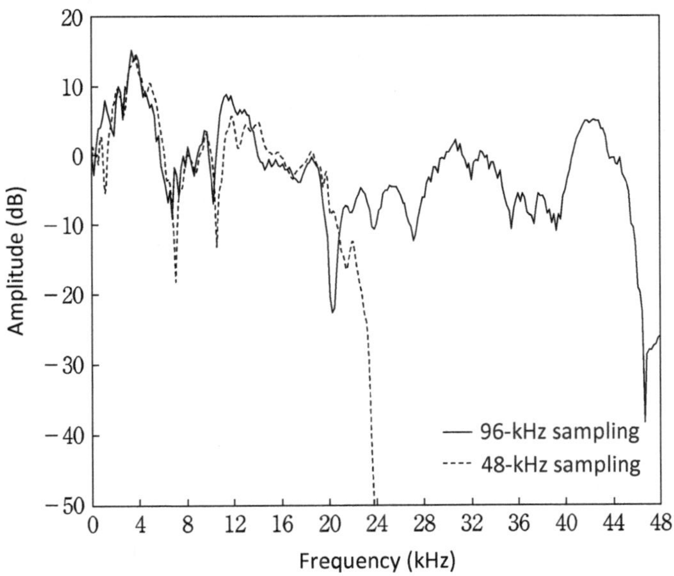

Fig. A.38 HRTFs of the front direction with sampling frequencies of 96 and 48 kHz

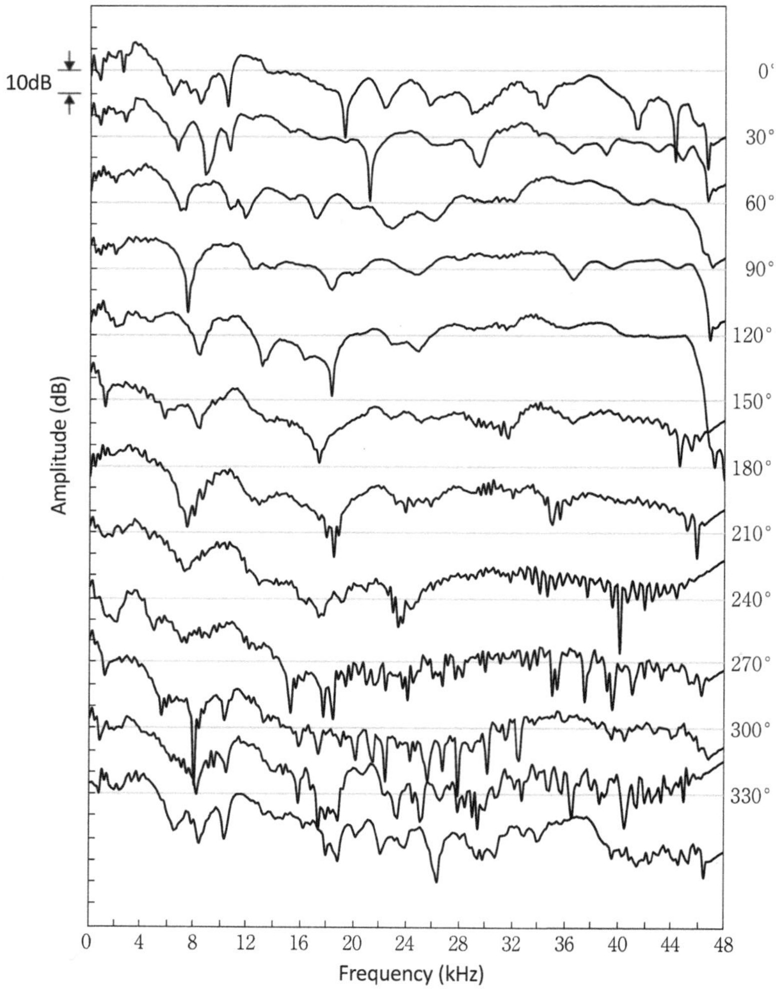

Fig. A.39 Amplitude spectra of HRTFs with 96-kHz sampling at the right ear for sound sources in horizontal plane (0°–330°, 30° steps)

References

Kurosawa A, Takagi T, Yamaguchi Z (1982) On transfer function of humanear and auditory localization. J Acoust Soc Jpn 38:145–151. in Japanese

Vorlander M (2011) Auralization:fundamentals of acoustics, modelling, simulation, algorithms and acoustic virtual reality. Springer

Index

K. Iida, *Head-Related Transfer Function and Acoustic Virtual Reality*,
https://doi.org/10.1007/978-981-13-9745-5